BHB

Chemically Modified Surfaces

Chemically Modified Surfaces

Edited by

Joseph J. Pesek
San Jose State University, California, USA

Ivan E. Leigh
CertainTeed Corporation, Blue Bell, Pennsylvania, USA

ROYAL
SOCIETY OF
CHEMISTRY

Chem
Sep/ae

Proceedings of a Symposium on Chemically Modified Surfaces
held in Malvern, Pennsylvania on 16–18 June 1993.

Special Publication No. 139

ISBN 0-85186-595-X

A catalogue record for this book is available from the British Library

Published by The Royal Society of Chemistry,
Thomas Graham House, Science Park, Cambridge
CB4 4WF

Printed in Great Britain by Bookcraft (Bath) Ltd.

Preface

The Fifth Symposium on Chemically Modified Surfaces, as with previous symposia in this series, provided a forum for the presentation of new scientific contributions on the chemical modification of different materials, surface characterization, and other topics of current interest. In addition to traditional topics for this series - methods for characterizing oxide surfaces and industrial applications of surface modifications - this symposium featured a plenary lecture on conversion of oxide to hydride surfaces, as well as sessions on modifying polymer surfaces, modifying and characterizing catalysts, and surface studies on membranes and thin films. Publication of this Proceedings of the Chemically Modified Surfaces Symposium enables dissemination to a wider audience of the work presented and discussed at CertainTeed Corporation's Technical Center in Blue Bell, Pennsylvania, June 16-18, 1993.

The organizers acknowledge the cooperation of the participants, particularly those who submitted manuscripts for publication in the Proceedings. Special thanks are extended to session chairpersons, and to CertainTeed Corporation for its support of the Symposium.

Joseph J. Pesek
Ivan E. Leigh

Contents

Opening Remarks

Good morning! Welcome to the fifth biennial Symposium on Chemically Modified Surfaces. And welcome to CertainTeed's Levecque Technical Center.

As noted in the fliers that most of you have received over the last months, this Symposium continues the tradition established by Ward Collins and Don Leyden of providing a forum for the presentation of current work on the modification and characterization of surfaces. The topics will vary from meeting to meeting, reflecting the diversity of research areas encompassed by the broader concept "surface chemistry".

Typically, an academic person coordinates the technical program, an industrial person organizes the Symposium. This year Dr. Joseph J. Pesek of San Jose State University arranged the program, and I, with the help of a lot of people here, and thanks to the corporate sponsorship of CertainTeed, have made the meeting arrangements.

As some of you know, it has been touch and go over the last couple of months as to whether we would even hold the Symposium, due to lower than expected registration in the light of a very tight economic climate. Historic Philadelphia is a fitting venue for this year's Symposium, since Joe Pesek and I have felt a little like our Founding Fathers did at Valley Forge in the Winter of 1777! However, it's spring-time again, and we've persevered! Thanks for hanging in there with us. Speaking of Founding Fathers, I'm delighted that Ward Collins is with us today. It's good to see other veterans of this Symposium, along with our newcomers.

I want to acknowledge the support of Maurice Kelley, Manager of Technical Services, and Domenic Tessari, Vice-President of the Insulation Group's R&D Division, for making it possible for the Symposium to come to pass this year. Although we had originally planned to hold it elsewhere, we know you will enjoy the meeting facilities here and your accommodations at the Guest Quarters in Plymouth Meeting. I want to thank Claire Miller, IG R&D's Administrative Supervisor, for coordinating all the arrangements here and at the hotel. Our site manager, Mary Chantry, has arranged for the auditorium facilities as well as for special food services. And many of you have talked with Carolyn Everhardt, who has taken reservations, put together the meeting folders, and handled the multitude of tasks that need to be attended to for a meeting of this kind. She will continue to be our resource person for the duration of the Symposium. Thanks also to Anita Fariello for her help, especially with faxes!

You should each have a meeting folder. In addition, writing supplies (pads of paper, pens and pencils) are available. In the folder you will find the meeting program and abstracts. We should have an attendance list later on - probably tomorrow. For those curious about what CertainTeed does, there's a "fact sheet" included. Also some fliers from VSP, who put out the Journal of Adhesion Science and Technology. Their books in honor of Drs. Pluedemann and Fowkes should be of particular interest to this group.

If you'll turn to the meeting program and abstracts fascicle now, I'd like to review the events coming up. These are on the yellow pages in that fascicle. For Wednesday - today - we'll be starting in shortly with the plenary lecture. For the remainder of the morning the topic will be "Modification of Polymer Surfaces". After lunch we'll spend the afternoon on "Membranes and Thin Films". There should be ample time for discussion of each presentation, and discussions can be continued informally at break time. Lunches will be held in the Whitpain Room, on the other side of the building. Mrs. Everhardt will be with us at breaks; let her know if you have any special needs. We have a board set up outside the auditorium for messages of a non-emergency nature, as well as sign-up sheets. Also, if you'd like to have dinner with a group of us at the hotel tonight (probably at 6:30), please sign up on that sheet by afternoon break.

Thursday's program includes papers on "Modification and Characterization of Catalysts" in the morning, followed by "Industrial Applications of Modified Surfaces" in the afternoon. If we finish up early, you may wish to go back to the hotel, but be sure to be back for happy hour and our Symposium dinner at six in the Whitpain Room!

On Friday we'll wrap up with the session on Characterizing Oxide Surfaces.

A limo back to the hotel will be available at the end of each meeting day.

It is now my great pleasure to introduce the program chairperson and plenary speaker for the Symposium, Dr. Joseph J. Pesek. Dr. Pesek took his Bachelor's degree in Chemistry from the University of Illinois, and his doctorate in analytical chemistry from UCLA. After teaching at Northern Illinois University, he moved to San Jose State University in 1978 and has been there ever since. He is currently Professor of Chemistry and

Chairperson of the Chemistry Department. From 1991 to the present he has been a Dreyfus Scholar at San Jose State. He has also been Visiting Professor at Ecole Polytechnique in Paris, and at the University of Aix-Marseilles. He was recently in France and Japan on sabbatical. I don't know if he had a chance to drop in on our St. Gobain parent company's R&D facility near Paris, but I can assure him that he would be most welcome to visit there.

Dr. Pesek is at home teaching both analytical and inorganic chemistry, and has given graduate courses in separation methods, NMR, and FTIR as well. Those in the audience who attended 1991's Symposium will recall that Dr. Pesek spoke on the synthesis, characterization, and modification of hydride silica surfaces. Today, his Plenary Lecture will be on Conversion of Oxide Surfaces to Hydride Surfaces. Dr. Pesek ---

Ivan E. Leigh
Blue Bell, Pennsylvania

Plenary Lecture
Conversion of Oxide Surfaces to Hydride Surfaces

Joseph J. Pesek
DEPARTMENT OF CHEMISTRY, SAN JOSE STATE UNIVERSITY,
SAN JOSE, CA 95192, USA

Oxides, particularly silica, being ubiquitous in nature have been used in many important chemical applications such as catalysts as well as separation media for various forms of chromatography[1,2]. In most applications the essential feature of the oxide material is its surface. The surface activity and chemistry of any oxide material is dominated by the presence of free hydroxide groups. While the OH groups often are necessary for certain applications, they can often be detrimental in others. Therefore, a reliable method that can substitute another chemical entity for hydroxide while retaining some form of reactivity should prove useful in a wide variety of applications where an oxide material is desirable.

One such approach for the modification of oxide surfaces involves the conversion of hydroxide groups to hydrides[3-8]. For silica this overall process can be described as follows:

$$Si-OH \quad ---> \quad Si-H$$

One method for accomplishing this conversion involves first chlorinating the surface with an appropriate reagent such as thionyl chloride:

$$Si-OH \quad + \quad SOCl_2 \quad ---> \quad SiCl \quad + \quad SO_2 \quad + \quad HCl$$

The halogenated surface can then be converted to the desired product via reduction by an inorganic hydride:

$$SiCl \quad + \quad LiAlH_4 \quad ---> \quad SiH \quad + \quad LiCl \quad + \quad AlH_3$$

Unfortunately this approach (chlorination/reduction) has some serious limitations. First, it is extremely sensitive to moisture in each step and requires appropriate precautions which are both time-consuming and labor-intensive. Second, an efficient condensation apparatus (at dry ice temperature or below) is required to trap the relatively volatile reduction byproducts.

Another synthetic method for the production of

hydride surfaces involves the controlled chemisorption of silanetriol, $HSi(OH)_3$, the hydrolysis product of a trisubstituted silane, on an oxide surface. The hydrolysis step can be described as follows:

$$HSiX_3 \quad + \quad 3H_2O \quad ---> \quad HSi(OH)_3 \quad + \quad 3HX$$

After hydrolysis, the silanetriol is attached to an oxide surface, such as silica, by a series of condensation reactions that also involves cross-linking to adjacent silanols to form siloxane linkages similar to those existing on the substrate surface. Under carefully controlled conditions, this process should result in a monolayer formation on the surface in which the hydroxides are replaced by hydrides:

$$\begin{array}{ccccc} & H & & & H \\ & | & & & | \\ -Si-OH & + & HO-Si- & ---> & -Si-O-Si- & + & H_2O \end{array}$$

The basic process involves the same methodology used in typical silane-coupling procedures[9,10]. Most often the hydride is replaced by an organic moiety and the reaction is then referred to as organosilanization. Organosilanization is the most frequently used reaction for modification of silica, and other oxide surfaces, used in chromatography, capillary electrophoresis, catalysis, and electrochemistry. Therefore, it is essential to evaluate the two approaches for the production of hydride surfaces as well as the ultimate usefulness of these materials as replacements for hydroxide surfaces.

There are several spectroscopic and thermal methods of analysis which can be used to characterize the hydride surface[5,6,8]. Among the best is diffuse reflectance infrared Fourier transform (DRIFT) spectroscopy. The Si-H stretching frequency for these materials is in the range of 2250-2260 cm^{-1}. This region of the infrared spectrum is generally free of interferences from other absorptions so it provides positive identification for the formation of the hydride surface. Some examples of the use of this method are given in Figure 1 for different sources of native silica. Each group of spectra is a comparison between the native material, the hydride formed from chlorination/reduction and the hydride formed from triethoxysilane (TES) silanization. For all the spectra where the hydride is present, a distinct strong band is seen at about 2259 cm^{-1} which represents the Si-H stretching absorption. In these spectra as well as those representing silica from other sources, a semi-quantitative evaluation of the extent of hydride formation can be made from the intensity of the Si-H stretching peak. In every case, the intensity of the peak for the hydride formed from the TES reaction is significantly greater than for the peak in the spectrum of the material formed in the chlorination/reduction process. Another feature of the DRIFT spectra which should be noted involves the isolated-silanol absorption

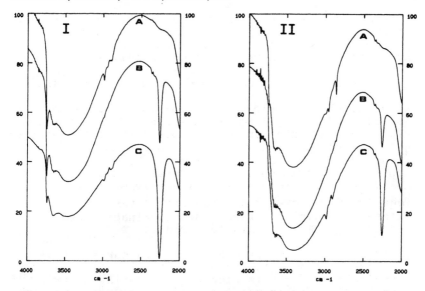

<u>Figure 1</u> Partial Drift Spectra of hydride modified silicas: (I) Partisil-40; (II) Vydac TP. Curves: (A) native silica; (B) chlorination/reduction product; (C) TES silanization product

band at 3739 cm^{-1}. Again in every case the intensity of this band is lower for the hydride material formed from the TES reaction. This is an important feature since isolated silanols are mainly responsible for both poor chromatographic and electrophoretic performance when certain solutes such as bases are separated by these methods.

Another useful spectroscopic method for characterizing modified surfaces is cross-polarization magic-angle spinning (CP-MAS) NMR. It is possible to make the same comparison using Si-29 CP-MAS NMR between the native silica, the chlorination/reduction product and the TES reaction product that was done by DRIFT. Figure 2 shows an example of such a comparison using the same batch of native silica. The peaks observed in the spectrum for the native material (A) are those which have been reported previously for many types of silica. They include the peak representing the siloxanes [Q_4, $Si^*(OSi\equiv)_4$] at -110 ppm, single silanols [Q_3, $HOSi^*(OSi\equiv)_3$] at -100 ppm and geminal silanols [Q_2, $(OH)_2Si^*(OSi\equiv)_2$] at -90 ppm. When the surface is modified by the chlorination/reduction procedure two additional peaks appear in the spectrum (B). These peaks are due to the $HSi^*-(OSi\equiv)_3$ resonance at -84 ppm and the $HSi^*(OH)(OSi\equiv)_2$ resonance at -74 ppm. These same two peaks also appear in the Si-29 CP-MAS NMR spectrum (C) of the hydride product from the TES reaction but are much more intense than in the chlorination/reduction spectrum. As in the DRIFT spectra, it can be concluded from the solid state NMR results that the TES reaction produces a

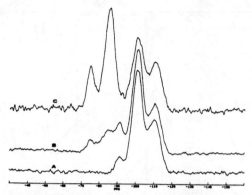

Figure 2   Si-29 CP-MAS NMr spectra of Partisil-40
silicas: (A) native silica; (B) chlorination/reduction
product; (C) TES silanization product

higher concentration of hydride on the surface than the
chlorination/reduction method.
    Further characterization of the hydride material is
possible through thermal methods of analysis. Both DSC
and TGA can be used to detect the following reaction of
the hydride material:

$$-Si-H \quad + \quad 1/2O_2 \quad ---> \quad -Si-OH \quad + \quad heat$$

Some examples of typical DSC curves from two sources of
native silica are shown in Figure 3. As in the

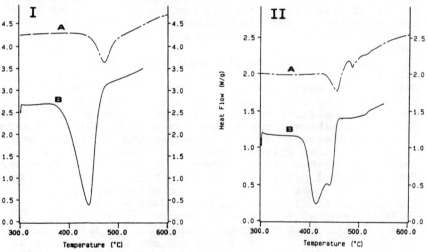

Figure 3   DSC curves of hydride-modified silicas in air:
(I) Partisil-40; (II) Vydac TP.   Curves: (A)
chlorination/reduction production; (B) TES silanization
product

spectroscopic methods, the DSC peak for the TES reaction
is larger than for the corresponding peak from the
chlorination/reduction method. There is a slight

displacement (20-30°) of the peak maximum to lower
temperature for the TES material. This shift may be due
to a slight excess over a monolayer (polymerization) on
the surface. For the Vydac material a second peak
appears in the DSC curve for both the chlorination/
reduction process and the TES reaction. The origin of
this second peak is not known but may be the result of a
more heterogeneous surface that affects the thermal
decomposition process. However, for both types of
silicas it is clear that the amount of hydride on the
surface is greater for the TES modification method.

Exact quantitative determination of the amount of
hydride on the surface can be done by placing the
material in a very strong base which will liberate
hydrogen gas[8]. The hydrogen gas can be collected and the
quantitative measurement is made by peak area
determinations from the gas chromatogram. With a good
standardization procedure, the precision of the
determination is about 2-4% relative standard deviation.
Knowing the surface area from BET measurements, the SiH
surface concentration can be calculated from the gas
chromatographic data. The original SiOH surface
concentration can be determined from TGA data on the
native material and a comparison between the two surface
concentrations, [SiH]/[SiOH] x 100, will then allow the
determination of the percent of silanols that were
converted to hydrides. Some results for these conversion
efficiencies on several types of silica using both
methods are shown in Table 1. It can be clearly seen
from the results in the table that the TES silanization
method is superior to the chlorination/reduction
procedure on all types of silica. Except for Vydac
silicas, this efficiency is generally close to 100% for
the TES product. The low values for the efficiency on
Vydac silica may be due to a material with many
micropores which are not measured by the BET method but
contain OH-groups resulting in a low surface area
measurement or perhaps strongly adsorbed water which
could lead to abnormally high surface-SiOH densities.
Either one or a combination of these two situations would
lead to a low efficiency determination.

The hydrogen displacement/gas chromatography method
(HD/GC) is a rather tedious and time-consuming procedure
for the determination of hydride content. Since the area

Table 1   Percent Hydride Modification Efficiencies

silica	chlorination/reduction product	TES silanization product
Partisil-40	24 (18)[b]	97
Partisil-10	23	79
Kromasil-100	28	126
Vydac TP (lot 1)	4	30
Vydac TP (lot 2)	6	35
Nucleosil 300-10	23	92

of the peak in the DSC thermogram is directly related to
amount of hydride on the surface, it should be possible
to estimate the SiH concentration from this data. A
comparison between the DSC peak area and the specific SiH
coverage determined from the HD/GC experiment is shown in
Figure 4. While there is a high degree of correlation

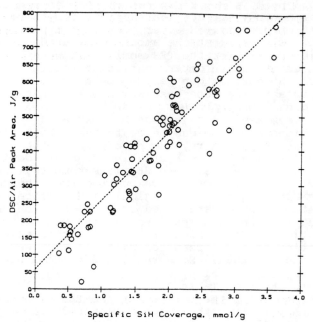

Figure 4 DSC/air peak as a function of surface coverage

between the two measurements, there is a considerable
degree of uncertainly associated with the DSC results
leading to the scatter in the data. Therefore, the DSC
measurement can be used to give a good estimate of the
SiH concentration on the surface but for an accurate
determination the HD/GC method must be used.

In order for the final product to have reproducible
properties and for the surface area and pore size to
remain the same, it is necessary to control the reaction
conditions of the TES silanization so that there is close
to a monolayer coverage[8]. If there is less than a
monolayer coverage, then a significant number of
unreacted silanols will remain. If there is more than a
monolayer coverage, then appreciable changes in the
surface area and perhaps more importantly the pore size
will occur. The primary factors in controlling the TES
reaction are solvent, temperature, acid catalyst
concentration and TES concentration. From solubility and
upper temperature limit considerations, it was determined
that dioxane was a suitable solvent for this reaction.
Variable temperature studies indicate that as expected
the rate of the reaction is accelerated by increasing the
temperature so that conditions near the boiling point of

the solvent result in rapid production of the hydride
surface. Similarly varying the acid concentration has
demonstrated that about 100 mM HCl leads to rapid
formation of the product without any significant
decomposition. Finally, the concentration of TES is
crucial in determining the extent of silanization on the
surface. Figure 5 shows the extent of SiH coverage with
respect to the concentration of TES for a one-hour
reaction time, a 100 mM concentration of HCl using
dixoane as the solvent at a temperature near the boiling
point (gentle reflux). In both types of silica shown in

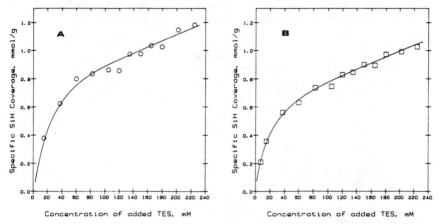

Figure 5　Specific SiH coverage as a function of TES
concentration: (A) Vydac TP; (B) Nucleosil 300-10

these examples there appears to be two parts to the
reaction curve. An initial rapid portion which reaches a
limiting value corresponding to about 8-10 micromoles/m^2
which is the expected amount expect for monolayer
coverage of hydride. The second slower portion of the
curve probably corresponds to formation of a multilayer
structure of the hydride polymer on the surface.
Therefore, from this data it can be concluded that a
concentration of about 100 mM TES will produce roughly a
monolayer of hydride on the surface in a one-hour
reaction time.

While the hydride surface is an interesting material
in itself, most applications result from using this
product as an intermediate for the attachment of various
organic moieties to the surface[6,7]. This can be
accomplished through a heterogeneous-phase hydrosilation
reaction such as the following:

$$Si-H \quad + \quad CH_2=CH-R \quad \xrightarrow{\text{cat}} \quad Si-CH_2-CH_2-R$$

The final product has a direct silicon-carbon linkage at
the surface which has been shown to be more
hydrolytically stable than the Si-O-Si-C linkage which
results from organosilanization. The only other approach
for making direct Si-C linkages involves two step

chlorination/Grignard or chlorination/organolithium reaction scheme[11]:

$$Si-OH \ + \ SOCl_2 \ ---> \ Si-Cl \ \xrightarrow{R-M} \ Si-R$$

Unfortunately this method has several drawbacks that limit its utility. The two-step halogenation/alkylation process is significantly more difficult than the one-step organosilanization procedure. The residual salts which result from the alkylation reaction can be occluded in the silica matrix and are difficult to remove. The chlorination step is extremely moisture sensitive and must be done under scrupulously dry conditions. Finally, the Grignard or organolithium reagent is subject to limitations on the number and kind of functional groups which may be present. This can severely limit the variety of bonded materials (R groups) which can ultimately be produced. Therefore, the ideal combination of the hydride intermediate produced via the TES silanization reaction followed by hydrosilation appears to be best approach for synthesizing bonded materials with direct silicon-carbon linkages at the surface.

A number of factors must be considered in order to optimize the hydrosilation reaction[6]. These include reaction temperature, reaction time, solvent, olefin concentration and choice of catalyst. The latter factor involves the greatest choice. There are literally hundreds of inorganic and organic complexes of transition metals such as platinum, palladium, rhodium, ruthenium, iridium and nickel which can successfully catalyze the hydrosilation reaction. Our own work has shown varying degrees of efficiency for several types of catalysts which depend to a certain degree on the olefin which is being bonded. One of the consistently most successful of these catalysts is the traditional Speier's catalyst, a 2-propanol solution of hexachloroplatinic acid, which has been used for many years in homogeneous phase hydrosilation reactions. It is certain that a systematic study of catalysts will be necessary to produce the highest yield of bonded materials as well as to avoid any contamination of the product through the reduction and/or deposition of catalyst on the surface. It is likely that no single catalyst will be found to yield the best product for every olefin which might be used.

With respect to the other factors which must be considered, a greater degree of progress has been made toward optimization. While the reaction proceeds without difficulty in a solvent such as toluene, it was found that whenever possible it is best to use the neat liquid depending on the physical state of the olefin at the reaction temperature. If a solvent is used, higher concentrations of the olefin lead to a higher bonding density on the surface. In fact there is roughly a linear relationship between olefin concentration and surface coverage. Figure 6 shows the effect of reaction time on the surface coverage for hydrosilation with 1-octene and 1-octadecene. It can be seen that the rate of

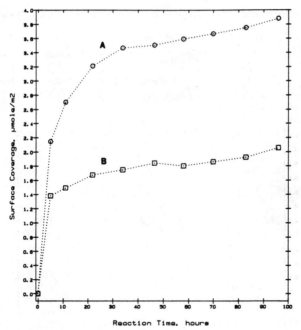

Reaction Time, hours

Figure 6 Effect of reaction time on alkyl surface coverage for the hydrosilation of (A) 1-octene and (B) 1-octadecene on hydride modified silica

the reaction is rapid at first and then small increases in coverage occur as the reaction proceeds. The difference in coverages between octyl and octadecyl is due to their differences in molar concentrations as neat liquids. This result is consistent with the concentration dependence observed when the olefins are placed in a solvent. The effect of temperature on the surface coverage for these same two olefins using a fixed reaction time is shown in Figure 7. In general the conclusion would be that a higher temperature would lead to a more rapid reaction and a higher yield. However, the highest temperature allowed is controlled by the boiling point of the liquid olefin. In addition, the thermal stabilities of the olefin, the product and the catalyst must also be considered. Higher temperatures have been shown to cause irreversible catalyst reduction which can be detected by a darkening of the product, presumably because the metal precipitates and is trapped inside the porous silica matrix. The possibility and extent of such a reduction depends on both the catalyst and olefin involved in the hydrosilation. Therefore, as observed in the efficiency studies, no single catalyst will be useful for all hydrosilation reactions.

Table 2 shows the alkyl surface coverages obtained when using a reaction temperature of 100°C for 60 h. The results of the hydrosilation can be compared to those obtained for organosilanization on the same batches of native silica. For C-8, the hydrosilation appears to

Table 2   Surface Coverages of Bonded Silica Phases

silica support	S_{BET}, m²/g	anchored group	n	M_R	% C	α_R, μmol/m²
Partisil-40	296.7	–C₈H₁₇	8	112.22	11.00	4.43
Partisil-40	315.3	–OSi(CH₃)₂C₈H₁₇	10	170.37	12.85	4.15
Partisil-40	296.7	–C₁₈H₃₇	18	252.49	13.40	2.48
Partisil-40	315.3	–OSi(CH₃)₂C₁₈H₃₇	20	310.64	20.45	3.67
Vydac 101TPB	89.1	–C₈H₁₇	8	112.22	2.10	2.51
Vydac 101TPB	88.8	–OSi(CH₃)₂C₈H₁₇	10	170.37	1.78	1.71

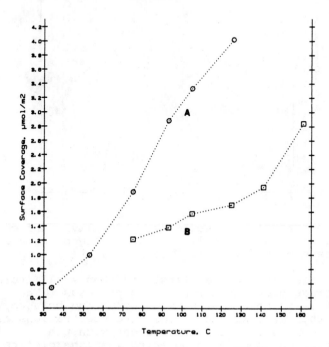

Figure 7 Surface coverage as a function of reaction temperature for hydrosilation of (A) 1-octene and (B) 1-octadecene on hydride silica

give better results under these conditions while for C-18 organosilanization appears better. However, these studies are quite limited and the results should be interpreted with caution since it is possible that not all reaction conditions have been optimized.

Spectroscopic studies of these materials are also useful in characterizing the nature of the products formed. Figure 8 shows the DRIFT spectra for a complete series of materials from the starting native silica, the hydride intermediate and two product phases. As discussed earlier, the Si-H peak is clearly visible near 2260 cm⁻¹ for the intermediate (B). After the hydrosilation reaction, the intensity of this peak diminishes and strong C-H stretching bands between 2800-3000 cm⁻¹ appear in the spectra (C & D) of the products. As both the reaction time and temperature studies have shown (Figures 5 and 6 as well as Table II), the extent of coverage is determined by the molecule which is to be

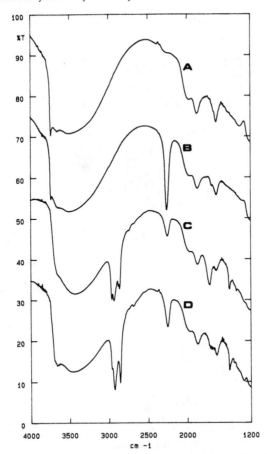

<u>Figure 8</u> Partial DRIFT spectra of silica after
derivatization: (A) native silica; (B) hydride
intermediate; (C) octyl-bonded phase; (D) octadecyl-
bonded phase

bound. Greater coverage is obtained for C-8 than C-18.
This can be seen in the DRIFT spectra shown in Figure 8
since the intensity of the Si-H peak is diminished to a
greater extent for C-8 than for C-18.
    Confirmation of the bonding process can be obtained
by Si-29 and C-13 CP-MAS NMR spectra. For example, the
Si-29 CP-MAS NMR spectrum of the hydride intermediate in
comparison to an octyl-bonded material is shown in Figure
9. For the hydride material (A), the peaks labeled d and
e represent silicon atoms with a hydride bound to them.
In comparison to the spectrum for the octyl phase (B),
these peaks have almost completely disappeared and two
new peaks (g and f) have appeared which represent
silicons with directly attached carbons. This spectrum
is perhaps the best direct evidence of the bonding
process since the peaks in the spectrum at -66.2 ppm and
-54.6 ppm can only appear if there is a direct silcon-

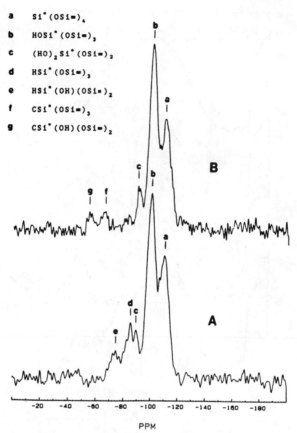

a   Si*(OSi≡)₄

b   HOSi*(OSi≡)₃

c   (HO)₂Si*(OSi≡)₂

d   HSi*(OSi≡)₃

e   HSi*(OH)(OSi≡)₂

f   CSi*(OSi≡)₃

g   CSi*(OH)(OSi≡)₂

Figure 9   Si-29 CP-MAS NMR spectra of (A) hydride
intermediate and (B) octyl bonded silica (Vydac TP)

carbon bond which results from the hydrosilation
reaction. The DRIFT spectrum provides indirect evidence
because of the diminishing of the Si-H stretching band
intensity. Additional indirect evidence is obtained from
the C-13 CP-MAS spectrum of the octyl phase shown in
Figure 10 The peaks in the spectrum can be readily
assigned according to the labels in the figure. The peak
at 12.2 ppm represents the methylene carbon directly
attach to the silicon at the surface. As seen in both
the Si-29 CP-MAS NMR and the DRIFT spectra, there are
unreacted hydrides which could potentially hydrolyze to
create chromatographically troublesome silanols.
Therefore an endcapping procedure to eliminate as many of
the accessible hydrides as possible would be useful. One
possible solution to this problem is the use of ethylene
gas since it represents the smallest possible olefin to
bond to the surface. Figure 11 shows the DRIFT spectrum
which results from bubbling ethylene gas through a
hydride silica/toluene suspension. The effects of
bonding can be readily seen by a diminishing of the Si-H
stretching frequency peak at 2260 cm^{-1} and the appearance

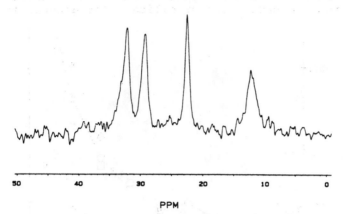

Figure 10 C-13 CP-MAS NMR spectrum of octyl bonded phase (Vydac TP)

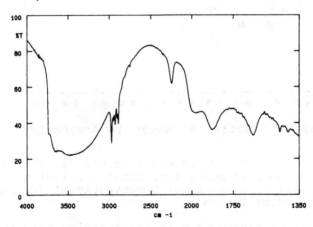

Figure 11 Partial DRIFT spectrum of hydride-modified silica reacted with ethylene gas

of C-H stretching bands between 2900 and 3000 cm^{-1} and two C-H bending bands at 1467 and 1418 cm^{-1}.

Chromatographic testing of the products as well as the hydride intermediate can be used to further characterize the materials. Of particular interest is the relative nature of the hydride surface in comparison to bare silica as well as alkyl-modified materials. Horvath[12] has suggested the use of crown ethers to determine the relative hydrophobicity or hydrophilicity of surfaces under varying mobile phase compositions. When the mobile phase is rich in water, then the hydrophobic nature of the surface can be probed with the

appropriate choice of solutes while for organic-rich
mobile phases the hydrophilic nature of the surface can
be examined. Figure 12 shows a plot of log k' for a
crown ether solute as a function of mobile phase
composition for several types of stationary phases all
made from the same batch of silica. The general shape of

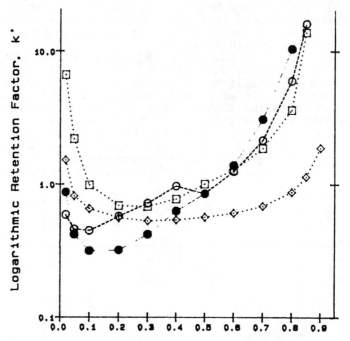

Volume Fraction of Water in Acetonitrile

__Figure 12__ Log k' for crown ether solute vs. volume
fraction of water in aqueous acetonitrile mobile phase:
<> native silica; ☐ hydride; O conventional C-8 phase;
● C-8 from hydrosilation.

the curve is in agreement with the behavior predicted by
Horvath with larger retention at both low and high
concentrations of water in the eluent. Significant
deviations from the mathematical model are most
pronounced in the range of 0.1 to 0.6 volume fraction of
water. Despite the lack of perfect agreement with the
theory, some interesting observations can be made from
the data in the figure. As expected, bare silica shows
the lowest retention when water is the major component of
the mobile phase due to its very hydrophilic surface.
Interestingly, the hydride and not bare silica shows the
greatest retention at low water concentration. This
would indicate some very active sites on the surface
which might be the result of partially condensed species
of the type =SiH(OH) from the silanization process.
Another interesting feature is that the hydride material

exhibits solvophobic interactions at high water
concentrations which is comparable to that of a
conventional C-8 phase. Because of the higher alkyl
content of the C-8 phase from the hydride, it is not
surprising that it displays the highest degree of
solvophobic interaction. More studies are needed to
completely explain the behavior of the hydride and the
alkyl phases produced from the hydride intermediate as
well as to further characterize the retention properties
at intermediate mobile phase compositions.

Another means of chromatographically probing the
nature of the hydride surface involves the use of basic
probes such as substituted anilines and comparing it to a
neutral species such as benzyl alcohol. Typically the
capacity factors for the anilines and the benzyl alcohol
are the same and low (k' ~ 0.4) at mobile phase
compositions such as 60:40 acetonitrile/water. The poor
selectivity results from a hydrophobic phase with a zero-
length alkyl chain. Figure 12 shows a typical
chromatogram for these solutes. In general the peaks

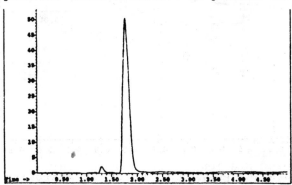

Figure 12 Elution profile for diethylaniline on hydride-
modified Vydac TP

are somewhat broad and asymmetric with retention times
for the anilines that often decrease upon subsequent
injections. These results indicate that the anilines are
interacting with a small number of highly active sites
such as the =SiH(OH) sites which might be present from
incomplete condensation during silanization. The
presence of these sites has been confirmed by Si-29 CP-
MAS NMR as shown in Figure 9 above.

Separation of a variety of solutes on C-8 and C-18
phases prepared from the hydride intermediate has also
been demonstrated. Of particular interest are difficult
species such as peptides, proteins and alkaloids. Figure
13 shows the separation of a five-component mixture of
barbital derivatives. Baseline separation of all the
sample components is achieved with excellent peak shape
in about 11 minutes.

Because the synthesis of alkyl-bonded phases via the
hydride intermediate produces a direct silicon-carbon

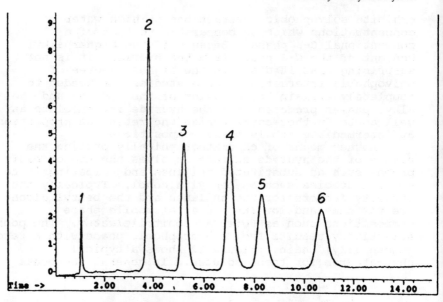

Figure 13 Separation of barbiturate mixture on C-18 phase prepared from hydride intermediate on YMC silica. Mobile phase 50:50 methanol water. Solutes: 1, impurity; 2, butabarbital; 3, amobarbital; 4, secobarbital; 5, phenobarbital; 6, hexobarbital

bond at the surface, the overall hydrolytic stability of these materials is of particular interest. In general silica-based stationary phases have limited stability in both acidic and basic solution. Figure 14 shows a comparison of the hydrolytic stability between a conventional C-8 phase from organosilanization vs. a C-8 phase prepared by hydrosilation in which the mobile phase contains 0.1% TFA at a pH of about 1.8. It can be clearly seen that the rate of degradation of the hydrosilation product is significantly less than that of the common commercial material. This result parallels those obtained with the material packed in a chromatographic column. In this case the differences are even more dramatic. After 3000 column volumes, the phase prepared through hydrosilation had lost less than 10% of its bonded material as measured by the retention of phenylheptane vs. almost 90% for the phase made by organosilanization. Limited testing of these materials has been done under basic conditions. Using 50 mM phosphate at pH = 9 for the mobile phase, retention, asymmetry. and peak efficiency have been measured for benzyl alcohol, dimethyl aniline and diethyl aniline. The anilines are especially sensitive to the presence of silanols which would be generated upon the cleavage of the bonded phase. Figure 15 shows the column efficiency for these probes on a C-18 phase prepared by

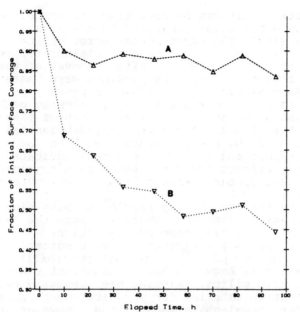

Figure 14 Relative surface coverage of C-8 phase on Vydac TP as a function of hydrolysis time. (A) C-8 prepared by hydrosilation; (B) C-8 prepared by organosilanization

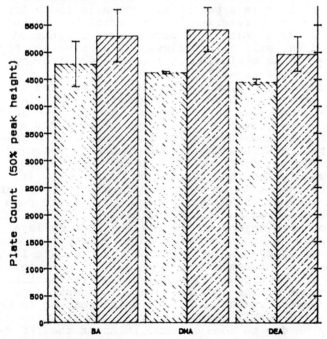

Figure 15 C-18 phase (hydrosilation) column efficiency before (left bar) and after treatment at pH 9. A, benzyl alcohol; DMA, dimethylaniline; DEA diethylaniline

hydrosilation. It can be seen that the efficiencies before and after being subjected to the alkaline mobile phase are within the experimental error of the measurement. Similar results are obtained for retention and asymmetry measurements. After removal of the material from the column the packings were tested by DRIFT. No peak for the stretching frequency of isolated silanols is detected and the Si-H stretching peak shows no apparent decrease in intensity. Both of these results indicate that there has not been any substantial degradation of the phase. Therefore, both the chromatographic and spectroscopic data indicate that no deleterious effects occur for short term exposure of bonded phases prepared through hydrosilation to alkaline conditions.

Despite the encouraging results at both high and low pH for phases made through hydrosilation, the use of silica will still have some pH limitations. Therefore, other materials may ultimately be more suitable than silica at extreme pH's [13]. One such possibility is alumina which is known to have an inherently higher pH stability than silica. Alumina has been used extensively as a separation material for purification purposes as well as for normal-phase separations. However, until recently there have been relatively few reports of alumina-based materials used in the reverse phase mode. Therefore, it seemed logical to try the same hydride/hydrosilation reaction scheme on alumina to determine if it is a potential synthetic route for the preparation of a reverse-phase material bonded to alumina [14]. In general, the same reaction protocol developed for silica is followed on alumina and the products can be evaluated by the same spectroscopic and thermal methods as well.

Figure 16 shows the DRIFT spectra for a series of alumina samples. In the spectrum of bare alumina (A), only the broad peak for adsorbed water between 3800 and 2600 cm^{-1} can be seen as a distinguishing feature. However, upon modification of the surface with TES, the distinct Si-H stretching frequency can be observed at 2260 cm^{-1} in the spectrum (B). Upon hydrosilation (spectrum C), the band at 2260 cm^{-1} diminishes in intensity and peaks in the C-H stretching region appear between 3000 and 2800 cm^{-1}. These results are parallel to those observed on silica indicating a direct bonding of the organic moiety to the silicon hydride layer.

Further evidence of the bonding process is obtained through Si-29 CP-MAS NMR spectroscopy. Figure 17 contains the spectra of the hydride alumina material (A) as well as the product of a hydrosilation reaction (B). The spectrum of the hydride material clearly shows the Si-H resonance at -85 ppm. It is interesting to note that small Q_4 and Q_3 peaks appear in the spectrum also. This may be due to some decomposition of the TES during the reaction or perhaps to greater than a monolayer coverage on the surface. After reacting an olefin with the hydride, a new peak at -65 ppm appears in the

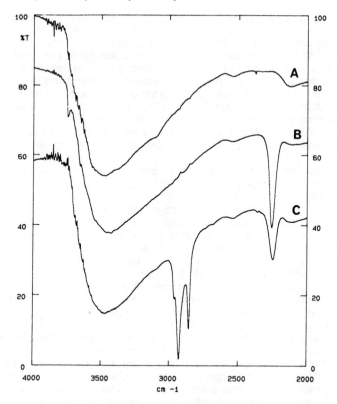

Figure 16 DRIFT spectra of alumina materials: (A) bare alumina, (B) hydride alumina, (C) reaction product of hydride alumina with 1-octadecene

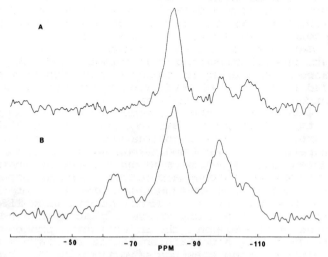

Figure 17 Si-29 CP-MAS NMR spectra of alumina materials: (A) hydride alumina and (B) reaction product of hydride alumina with 1-octadecene

spectrum which corresponds to the $(OSi\equiv)_3Si^*-C$ species, indicating successful bonding to the surface. The Al-27 CP-MAS spectrum also provides evidence of bonding to the surface. The spectrum consists of two peaks: one which represents the tetrahedrally coordinated aluminum ions on the surface (~20% of the sites) and the second which represents the octahedrally coordinated ions (~80% of the sites). Upon formation of the hydride, the octahedral peak undergoes distortion probably as a result of increased quadrupolar effects. The distortion increases further upon subsequent reaction of the hydride with an olefin.

Thermal analysis also confirms the bonding of the hydride to the alumina. The DSC/air curve has a maximum between 520-530°C. This value can be compared to the pure polymeric material, polyhydrosiloxane (polymerization product of TES in the absence of alumina), which has a maximum at 365°C and for the hydride silica which has a peak maximum at 430-450°C. The large shift in the peak for the thermooxidative process indicates that the hydride is chemically bonded to the surface as opposed to being chemically adsorbed and the larger shift in comparison to silica indicates a greater thermal stability for the alumina-based material.

In principle, the process which has been thoroughly characterized for silica and studied significantly on alumina should apply to any oxide surface. Preliminary studies have also been initiated on zirconia, titania and thoria. The results from the DRIFT and Si-29 CP-MAS NMR data indicate that the same conclusions drawn about the TES reaction on alumina can be made on these other oxides. Bonding of the TES to the oxide surface occurs readily forming a silicon hydride intermediate. This intermediate reacts with a variety of terminal olefin compounds to form an alkyl-modified surface. While the chemistry of these reactions can be directly adapted from the protocols established on silica, further characterization of all these materials is necessary. In particular, the chromatographic performance of stationary phases based on oxides other than silica which have been synthesized by the hydride/hydrosilation method must undergo extensive evaluation.

Because the chemistry on a capillary surface is identical to that of porous silica, silanols are the reactive groups, the same reaction schemes described above can be used for wall modifications in capillary electrophoresis. The method must be altered so that the reagents are passed through a capillary tube as opposed to the typical reaction vessels which are used for particulate silica. First a solution containing TES is passed through the capillary which is placed in a heated oven in order to accelerate the silanization process. Then a second solution containing the appropriate terminal olefin in the presence of a transition metal catalyst is pumped through the capillary. As in the case of oxide materials, all of the experimental conditions for both reactions must be optimized so that the capillary wall will be as completely derivatized with as few accessible silanols as possible.

Unfortunately, the spectroscopic and thermal analysis techniques which are available to characterize porous oxide materials cannot be used to evaluate reactions on capillary walls. This is due to the low surface area which leads to a very small total amount of material that is actually on the wall. For all of the analytical methods described above, this quantity of bonded material is below the detection limit.

Therefore, the only possible way to evaluate whether or not the bonding process has been successful is through the use of carefully selected electrophoretic experiments. Because the unmodified wall of a capillary will strongly adsorbed many solutes, particularly bases, it is possible to test the effective surface coverage by attempting HPCE separations of basic compounds at pHs where strong adsorption is expected. Figure 18 shows a separation of basic proteins on a capillary modified through TES silanization followed by hydrosilation. The

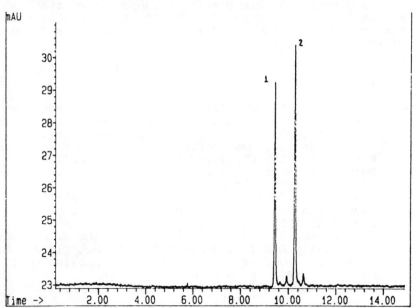

Figure 18 HPCE separation of turkey (1) and chicken (2) lysozymes in lactic acid/alanine buffer at pH 3.74

olefin used in this example contains an epoxide functional group that has been subsequently hydrolyzed to form the diol. For the same buffer conditions in the absence of capillary modification, the proteins are not eluted from the column but are irreversibly adsorbed on the walls. A very similar separation can be obtained on the hydride modified capillary alone. Here the silanols are effectively replaced by hydrides so that any interactions with basic solutes are eliminated. In both cases the electroosmotic flow is close to zero since the ionizable silanols have been virtually eliminated.

In conclusion, the silanization/hydrosilation method appears to offer promise in forming modified oxide surfaces for chromatography and capillary electrophoresis. This same procedure can potentially be used for other applications such as catalysis, solid-phase extraction and gas purification among others. The extent of surface coverage is excellent when compared to organosilanization and the product seems to offer good hydrolytic stability. The process is simple and is applicable to the attachment of virtually any organic moiety to the oxide surface. Further improvements in the synthetic method can be expected as well as more thorough characterization of the products. Finally, applications which will make use of the unique features of these materials must still be identified.

ACKNOWLEDGEMENT

This work has been partially supported by the National Science Foundation (CHE 9119933), the Petroleum Research Fund, the Dreyfus Foundation through a Scholar grant and The Separations Group.

REFERENCES

1. K.K. Unger, "Porous Silica-Its Properties and Use as Support in Column Liquid Chromatography", Journal of Chromatography Library, Vol. 16, Elsevier Scientific Publishing Co., New York, 1979.
2. R.K. Iler, "The Chemistry of Silica-Solubility, Polymerization, Colloid and Surface Properties, and Biochemistry", John Wiley & Sons, New York, 1979.
3. J.J. Pesek, "Chemically Modified Surfaces", D.E. Leyden and W.T. Collins, eds., Gordon and Breach New York, 1990, p. 93.
4. J.J. Pesek, J.E. Sandoval, C.-H. Chu and E. Jonsson "Chemically Modified Surfaces", H.A. Mottola and J.R. Steinmetz, eds., Elsevier Scientific Publishing Co., New York, 1992, p. 57.
5. J.E. Sandoval and J.J. Pesek, Anal. Chem., 1989, 61, 2067.
6. J.E. Sandoval and J.J. Pesek, Anal. Chem., 1991, 63, 2634.
7. J.E. Sandoval and J.J. Pesek, U.S. Pat. 5017540, 1991.
8. C.-H. Chu, E. Jonsson, M. Auvinen, J.J. Pesek and J.E. Sandoval, Anal. Chem., 1993, 65, 808.
9. E.P. Plueddemann, "Silane Coupling Reagents", Plenum Press, New York, 1982.
10. L.C. Sander and S.A. Wise, Anal. Chem., 1984, 56, 504.
11. J.J. Pesek and S.A. Swedberg, J. Chromatogr., 1986, 361, 83.

12. K.E. Bij, C. Horvath, W.R. Melander and A. Nahum, <u>J. Chromatogr.</u>, 1981, <u>203</u>, 65.
13. J.A. Blackwell, <u>J. Chromatogr.</u>, 1991, <u>549</u>, 59.
14. J.J. Pesek, J.E. Sandoval and M.G. Su, <u>J. Chromatogr.</u>, 1993, <u>630</u>, 95.

New Synthetic Methodology for Grafting at Polymer Surfaces

David E. Bergbreiter

DEPARTMENT OF CHEMISTRY, TEXAS A&M UNIVERSITY, COLLEGE
STATION, TX 77843–3255, USA

1 INTRODUCTION

Modification of a polymer's surface has developed into
an attractive way to chemically alter a polymer. Such
synthetic approaches potentially can maintain a
polymer's desirable bulk properties but can provide new,
different interfacial properties. Interfacial properties are of well recognized importance in many areas.
For example, biocompatibility, adhesion, permeability
and wettability are all important technological areas
that a polymer's surface chemistry can impact.[1,2]

There are many methods by which polymer surfaces
can be chemically modified. Most common examples of
this chemistry are reactions that introduce a single
type of functional group or mixture of functional
groups. These functional groups can be introduced onto
a surface with varying degrees of surface selectivity
and with varying extents of two dimensional randomness.
Methods by which surface selectivity can be controlled
and the sorts of methodology available to introduce single functional groups have both been reviewed.[3-5] This
paper briefly summarizes some existing as well as some
newer examples from our laboratory of an alternative
chemistry - surface grafting - which introduces multiple
numbers of functional groups and a distinct covalently
bound second polymer phase.

2 SIMPLE POLYMER SURFACE FUNCTIONALIZATION

In either simple functionalization or grafting
chemistry, the underlying reactivity of a substrate
polymer must be expressly considered. As illustrated in
Figure 1 polymer substrates vary considerably in their
reactivity. Unreactive polymers like polyethylene in
principle only contain primary, secondary and tertiary
C-H bonds and C-C bonds. Neither type of functional
group is typically considered to be a "reactive" functional group. Thus, harsh oxidizing reagents that lead

to chain scission and etching of substrate polymers are typically used. While the activity of such reagents can be modified by controlling exposure time or by

polyethylene	poly(ethylene terephthalate)	polyvinylchloride
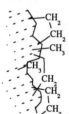	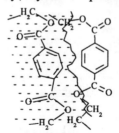	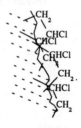
only C-H and C-C bonds which are chemically rather inert are present	reactive carboxylic ester groups are present which can react to form other carboxylic acid/alcohol derivatives	elimination easily occurs but forms a more chemically reactive surface which undergoes various side-reactions

Figure 1. Examples of polymer substrates with different reactivities.

controlling reaction conditions in reactions like gas phase plasma treatments, some physical modification of the surface is difficult to avoid. In the case of other unreactive polymers like poly(tetrafluoroethylene) or poly(chlorotrifluoroethylene), some of the more success-ful surface modification procedures rely on rather non-selective reductions which generate graphitic or poly-acetylene-like surface layers.[6] Controlled modification of more reactive ester-containing polymers is generally more easily accomplished. Several examples in the lit-erature detail methods to control the depth and chemical character of the product surfaces.[7-9] Finally, some polymers pose additional problems in that their initial reaction products are more reactive than the starting polymer. Poly(vinyl chloride) is a classic example of this situation. Elimination of H-Cl from this polymer produces a more reactive substrate and it is difficult to avoid formation of polyunsaturated colored surface layers in derivatization of PVC.

3   SURFACE GRAFTING

While simple surface modification of functional groups is one route by which chemistry at a surface can be controlled, there is also interest in processes that lead to surface grafting.[10] Grafting in polymer chemis-try is a process by which side chains of a second poly-mer are introduced onto an existing polymer main chain. Surface grafting generally leads to an overlayer of a second functionalized polymer covalently linked to the

substrate polymer. Unlike simple functionalization
which produces a polymer surface or interphase with an
intimate mixture of functional groups of the substrate
polymer and the newly introduced groups, grafting produces a physically distinct overlayer with properties
that resemble those of the pure graft homopolymer. Indeed, some of the so-called simple processes mentioned
above (PTFE reduction, PVC modification with base) actually produce surfaces that resemble the grafted surfaces
discussed below.

Grafting Chemistry Background

While surface layers of modified material can be
prepared using simple functionalization chemistry, surface grafting is really a polymerization that incorporates additional monomer from solution to form a second
covalently bonded surface polymer phase. However, this
type of grafting is not true grafting in that a true
graft involves modification of each polymer molecule.
The chemistry discussed below deals with this former
process of surface grafting. In this chemistry, an existing solid polymer has to first be chemically modified
in some fashion so that there are reactive sites at the
polymer's surface. These sites are then used to initiate graft polymerization or are coupled to some macro-
monomer. The product polymer contains a mixture of
polymer chains. Those polymer chains (indeed most of
the polymer chains) in the substrate that do not extend
to the surface are unmodified in this chemistry. Thus,
most of the substrate polymer remains unmodified. However, a second oligomer or polymer is covalently attached to some chains of the substrate polymer at the
surface.

Existing methodology for grafting is based predomi-
nately on radical polymerizations. Anionic graft polymerizations, cationic graft polymerizations and coordination graft polymerizations are less common. In the discussion below, we have focused our attention on a particular substrate polymer - polyethylene. The discussion begins with a brief summary of existing chemistry
relevant to surface graft polymerization on polyethylene. This summary is then followed by examples of chemistry we have developed and are developing as alterna-
tives to this existing synthetic graft chemistry.

Radiation induced grafting as illustrated in Figure
2 is a very general procedure applicable to all polymers.[11-13] While radicals generated by exposure to ion-
izing radiation from a source (^{60}Co, for example) are
generated throughout the polymer sample, surface radicals are most accessible and can either directly be used
in polymerization of a polymerizable monomer or can be
trapped as peroxides by oxygen in air. Subsequent exposure to heat or light can then generate a hydroxyl and
polymer-bound alkoxyl radical which then lead to polym-

erization. While such procedures are very general, they are not synthetically sophisticated. They afford little control over the amount and distribution of initiation sites. Homopolymerization of the vinyl monomer induced by soluble radicals formed in the peroxide decomposition also unproductively consume monomer.

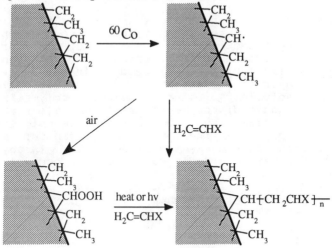

Figure 2. Radiation induced radical formation from polymer like polyethylene with subsequent direct grafting or grafting via an intermediate peroxide.

Radical sites at surfaces can also be generated by hydrogen-transfer to soluble radicals.[14] This can occur when a photogenerated radical (e.g. a benzophenone triplet) is added to a substrate polymer (Figure 3) or

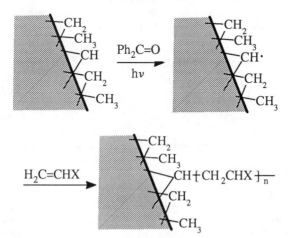

Figure 3. Grafting via photoinitiated chain transfer radical formation.

can occur when a substrate polymer is present during a
conventional radical polymerization.   As was true in
radiation-induced grafting, little control over graft
size and distribution is possible.   However, extensive
grafting can be obtained using either this procedure or
the radiation chemistry of Figure 1.

Several less conventional approaches to grafting
illustrated in Figures 4 and 5 provide good examples of
alternative procedures.   In the first of these procedures, the hydrophobic interaction of polyethylene with
an amphiphilic substrate was used to initially self as-
semble a layer of surfactant.[15] Polymerization initiated by a soluble radical then led to a polymerized
layer.   Though a direct mechanism for covalent attachment of this graft layer to the substrate polymer was
not provided, it is likely that chain transfer could
have occurred to generate covalent links.   In any case,
the graft layer reportedly was stable in contact with
solvents.

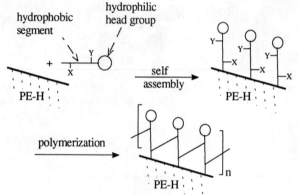

Figure 4.   Grafting of physisorbed layers.

A second example of an alternative procedure
(Figure 5) is work by Wnek's group which provided a
route to a polyethylene-polyacetylene composite.[16]   In
this chemistry, a Ziegler-type catalyst for polyacetylene formation was first incorporated into a swollen
polyethylene substrate.   Catalyst activation and acetyl-
ene polymerization then produced a polyacetylene layer
on the substrate polymer.

New Grafting Chemistry

We have explored several new synthetic approaches
to grafting onto polyethylene surfaces.   The first two
of these have relied on new chemistry for generation of
radical sites on polyethylene.   The other approaches use
an entirely different concept.   Each of these schemes is
discussed in more detail below.

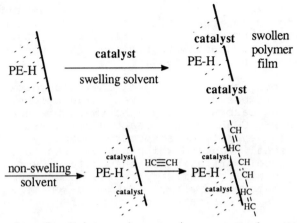

<u>Figure 5</u>. Grafting using an impregnated catalyst.

As noted above, there exist a number of approaches for introduction of functional groups onto polymer surfaces.[1,3] In the case of hydrocarbon polymers like polyethylene, such chemistry usually relies on oxidation reactions. For example, oxidative plasma treatments or $CrO_3 \cdot H_2SO_4$ etching product a mixture of carboxyl and carbonyl groups at the surface of polyethylene films. Conversion of carboxyl groups into potential initiator groups for free radical polymerization via formation of peroxides is a known method for transforming this initial functionalized surface into a grafted polymer. However, peroxides generate both a soluble and an insoluble radical and lead to unproductive hompolymerization. We recently reported an alternative to this chemistry that uses thiohydroxamic acid esters.[17] These esters, prepared as shown in equation 1, can be thermolyzed or photolyzed to

$$\text{(1)}$$

generate a polymer-bound alkyl radical and a soluble thiopyridyl radical. The latter radical is not very reactive and apparently does not lead to polymerization of monomers like acrylonitrile or methacrylonitrile under our conditions. However, the polymer-bound radical is competent and does initiate graft polymerization. The

extent of graft polymerization in this instance varies
depending on the concentration of monomer, temperature
and solvent because of the ability of thiohydroxamic
acid esters to react with radicals in a chain-terminat-
ing chain transfer process.

An important advantage of the thiohydroxamic acid
grafting process using oxidized polyethylene, PE-[CO_2H],
is that the carbonyl groups that were present in the
starting PE-[CO_2H] do not react under these conditions.
Therefore they provide a spectroscopically distinguish-
able internal standard which can be used to assay the
degree of polymerization of the graft. Typical polyethylene grafting chemistry relied on mass changes in
the film before and after grafting or on more qualitative assays to estimate the extent of grafting.

A disadvantage of the thiohydroxamic acid process
was that it is a multistep process relying initially on
a chain scissive oxidation. To circumvent this problem,
we have recently begun to study an alternative process
that does not depend on a prior functionalization reaction. The process we have developed relies on the fact
that residual unsaturation is present in polyolefins.
This residual unsaturation is readily detected by IR
spectroscopy in the form of a small peak at 908 cm^{-1}
(Figure 6a). In developing this chemistry, we relied on
several known reactions. First, the reaction of carbon-
carbon double bonds with B-H bonds is a well established
rapid reaction in synthetic organic chemistry.[18] In addition, recent studies have rediscovered the fact that
trialkylboranes can usefully react with oxygen to generate radicals which are useful in initiation of radical
chemistry.[19] Given that Chung's group has recently described the utility of polymer-bound alkylboranes as initiators for graft polymerization,[20] we were hopeful
that similar chemistry could be equally useful in grafting onto a solid polymer like polyethylene that contained residual alkene groups.

The expectation that the residual carbon-carbon
double bonds of polyethylene could be hydroborated was
tested by treatment of polyethylene films with THF solutions of $BH_3 \cdot SMe_2$. Subsequent oxidation with alkaline
hydrogen peroxide and trifluoroacetylation with tri-
fluoroacetic anhydride was then used to introduce a
spectroscopically detectable group. As expected, IR
spectroscopy showed that this chemistry was successful.
The product films from this reaction sequence either
lacked or contained C=C peaks of greatly diminished intensity. In addition, the product films contained a
readily detectable -$OCOCF_3$ group at ca. 1780 cm^{-1}. XPS
spectroscopy confirmed this result showing an increase
in the O_{1s} and F_{1s} signals relative to the C_{1s} signal
after completion of this reaction sequence.

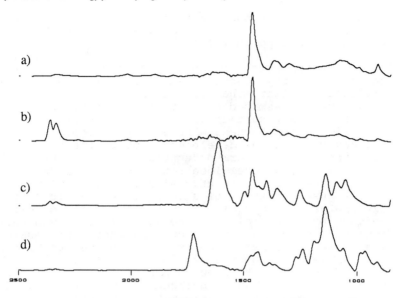

Figure 6. ATR-IR spectra of polyethylene films grafted
with vinyl monomers after successive treatment with
BH$_3$·SMe$_2$ and a vinyl monomer/air: (a), starting polyethylene film; (b), film grafted with acrylonitrile; (c)
film grafted with N,N-dimethylacrylamide; (d) film
grafted with methyl methacrylate.

We achieved varying success in applying this chemistry to polyolefin films as is shown by the results in
Figure 6.[21] Our best results were obtained using
acrylic acid derivatives as monomers. For example, as
shown by comparison of Figure 6a with Figure 6c or Figure 6d, polymerization of either N,N-dimethylacrylamide
or methyl acrylate was quite successful. In both cases,
a substantial C=O peak is evident in the product film
and the peak due to the residual C=C has largely disappeared. However, in the case of the monomer H$_2$C=CHCN,
the only obvious evident of a chemical change from Figure 6b is the disappearance of the C=C peak. In this
latter case, XPS spectroscopy did successfully show a
significant incorporation of nitrogen in the top 50 angstroms of the film. Close examination of the IR spectrum 6b also shows a small peak for the nitrile group.

While the graft polymerizations using an intermediate borane are successful with acrylic acid derivatives,
we have not as yet been able to attain high surface
selectivity. This is most clearly evident in the case
of the methyl acrylate grafting. In this case, the
starting polymer film approximately doubles in mass and
the tensile properties are obviously different after the
graft polymerization. Specifically, the starting polyethylene film is a flexible film while the product is a
stiff, hard, inflexible film. The film is also visually

thicker after the polymerization. We believe this reflects initiation of methyl acrylate polymerization
throughout the polymer film.

Both the thiohydroxamic acid chemistry and the
borane chemistry discussed above are examples of new
grafting chemistry that are clearly patterned after
known approaches used to initiate grafting onto polymers
or onto polymer surfaces. We are also exploring less
traditional approaches. These untraditional approaches
all derive from our developments of a type of functionalization we have termed "entrapment functionalization".[22] The general idea for this process derived from
the concept that a "functionalized" polyethylene is
really just a mixture of a small amount of functionalized polyethylene chains mixed with a large excess of
unfunctionalized polyethylene molecules. Given this
scenario, we envisioned a retro-synthetic process like
that shown in Scheme 1 below in which the functionalized
polyethylene would be disassembled into its functionalized and unfunctionalized components. In this way,
preparation of a functionalized polymer was reduced to
the development of suitable methods to prepare terminally functionalized oligomers and to the development
of a way to suitably combine these functionalized and
unfunctionalized components.

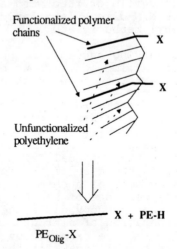

Scheme 1. Dissection of a functionalized polyethylene
into a functionalized component (PE_{Olig}-X) and an unfunctionalized component (the host polymer PE-H).

Entrapment functionalization is a process that is
essentially a blending process. In a typical procedure,
0.5-2.0 g of the terminally functionalized ethylene
oligomer is added to 100 g of virgin polyethylene. This
solid mixture is then suspended in a good solvent for
polyethylene. This solvent can be varied. Typical examples would include dibutyl ether, toluene and *ortho-*

dichlorobenzene. Subsequent heating of these suspensions produces a solution of the oligomer and the host polymer. Cooling this solution then produces a polyethylene precipitate. Alternatively, solution casting produces a film. Our work has shown that the terminal functional groups of these oligomers are largely or exclusively located at the surface of the product polyethylene films and powders that result from this blending process.[22-25] This scheme is diagrammed in Scheme 2.

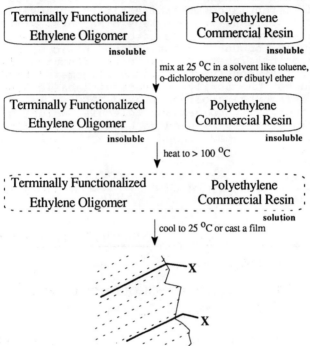

Scheme 2. Entrapment functionalization steps in which an oligomer (PE_olig-X) is first co-dissolved with a host polymer and then co-precipitated as a surface-functionalized powder or co-cast as a surface-functionalized film.

One of the most versatile features of the entrapment functionalization process shown in Scheme 2 is the ease with which different functional groups can be incorporated at the terminus of the ethylene oligomer and, hence, into the product functionalized polymer.[26] While much of our original work focused on using this synthetic versatility to incorporate a variety of spectroscopic labels, we have now begun to explore the potential of this chemistry in grafting. Three specific examples of this chemistry are discussed below. First, the potential for synthesis and characterization of a diblock oligomer as a reagent that incorporates a graft oligomer into polyethylene will be discussed. Ex-

tensions of this concept to incorporate thermally stable
radical initiators will next be described. Finally, the
applicability of this strategy for incorporation of anionic macroinitiators for anionic grafting onto polyethylene surfaces will be detailed.

    Our initial effort to use entrapment functionaliza-
tion for grafting began with the synthesis of a ethyl-
ene-poly(ethylene glycol) diblock oligomer. This synthesis which is shown in equation 2 had several features
which differentiated it from earlier syntheses of terminally functionalized ethylene oligomers. First, the
we were able to purify the intermediate carboxylic acid,
$PE_{Olig}-CO_2H$ through adsorption onto basic alumina. This
chemisorption process separated the functionalized
oligomer from non-functionalized oligomer $(CH_3(CH_2)_nCH_3)$
that did not chemically interact with basic alumina.
Second, since the cooligomer used (PEG_{Olig}) was available commercially in various molecular weight grades, [1]H
NMR could directly be used to both verify the success of
the synthetic procedure and to directly assay the PE_{Olig}
molecular weight.

$$H_2C=CH_2 \xrightarrow[\text{TMEDA}]{\text{BuLi}} \underset{PE_{Olig}\text{-Li}}{CH_3(CH_2CH_2)_n CH_2Li} \xrightarrow[\text{2. } H_3O+]{\text{1. } CO_2} \underset{PE_{Olig}\text{-CO}_2H}{CH_3(CH_2CH_2)_n CH_2CO_2H}$$

$$\xrightarrow{SOCl_2} \underset{PE_{Olig}\text{-COCl}}{CH_3(CH_2CH_2)_n CH_2COCl} \xrightarrow{HO(CH_2CH_2O)_n CH_2CH_2OCH_3}$$

$$(2)$$

$$\underset{PE_{Olig}\text{-PEG}_{Olig}\text{-OCH}_3}{CH_3(CH_2CH_2)_n CH_2CO_2(CH_2CH_2O)_n CH_2CH_2OCH_3}$$

    Using these procedures, a range of $PE_{Olig}-PEG_{Olig}$
diblock cooligomers with varying sized PEG groups could
be prepared. The PEG groups used ranged in molecular
weight from 350 - 4000. All of these diblock
cooligomers were entrapped in polyethylene. They in
general exhibited modest-good surface selectivity. XPS
spectroscopy, contact angle measurements and ATR-IR
spectroscopy all showed that the larger PEG groups led
to the best surface selectivity.[27]

    This entrapment chemistry which directly led to
formation of more hydrophilic polyethylene films with
PEG_{Olig} groups at the surface has most recently led to a
very different approach to grafting.[28] This approach
relies on entrapment functionalization and the enhanced
surface selectivity for oligomer entrapment engendered
in a polyethylene oligomer by incorporation of a poly-
ethylene-incompatible poly(ethylene glycol) block.
Using the synthetic chemistry in equation 3, we prepared
a monofunctionalized poly(ethylene glycol) oligomer with
a thermally stable trihaloacetate group as an end group.

$$HO-(CH_2CH_2O)_{\overline{n}}CH_2CH_2OH \xrightarrow{ClCOCCl_3} HO-(CH_2CH_2O)_{\overline{n}}CH_2CH_2OCOCCl_3$$

$$\xrightarrow{PE_{Olig}\text{-}COCl} PE_{Olig}\text{-}CO_2-(CH_2CH_2O)_{\overline{n}}CH_2CH_2OCOCCl_3 \tag{3}$$

Subsequent esterification with PE_{Olig}-COCl then led to a diblock oligomer containing a trichloroacetate as a potential initiator.[29]

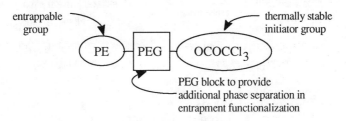

Once trichloroacetate terminated PEG groups were selectively entrapped at the surface of polyethylene films, a subsequent treatment with $Mn_2(CO)_{10}$ and light could be successfully used to generate a polyethylene-bound dichloroacetate radical which reacted vinyl monomers like methacrylonitrile. In a typical procedure, a 0.7 weight percent loaded film was suspended in DMSO with 0.7 mM $Mn_2(CO)_{10}$. Irradiation with a 150 W lamp at 25 °C was carried out for 2 h. The product polymer was then removed from the reaction mixture and extracted in a Soxhlet apparatus with acetone (a good solvent for poly(methacrylonitrile). The product film was then dried and analyzed by ATR-IR spectroscopy, XPS spectroscopy and by contact angle analysis using doubly distilled water as the test fluid.

As shown in Table 1, varying degrees of polymerization of the graft could be obtained. The degree of polymerization in this case was measured based on the N_{1s}/Cl_{2p} ratio and is probably a minimal estimate since it assumes every XPS detectable trichloroacetate initiated a graft polymerization.

Table 1. Radical Grafting of Methacrylonitrile onto a
        Polyethylene Film Containing Polyethylene-
        Poly(ethylene glycol) trichloroacetate Groups
        as Macroinitiators.

Time (h)	Monomer (M)	Degree of Polymerization	$c_{N_{1s}}/c_{Cl_{2p}}$	Contact Angle (°)
0	4.0	0	0.0016	110
2	5.5	16	0.013	–
2	3.5	10	0.011	84
2	1.5	6	0.007	87

The radical grafting described above was extended using a precursor of an anionic macroinitiator prepared according to equation 5. This oligomer contained a weakly acidic proton in the form of a terminal diphenylmethyl group.

$$CH_3CH_2CH_2CH_2Li \xrightarrow[\text{TMEDA, heptane}]{CH_2=CH_2,\ 30\ psig} CH_3CH_2(CH_2CH_2)nCH_2CH_2Li \xrightarrow[\text{2. } CH_3OH]{\text{1. } H_2C=CPh_2}$$

$$CH_3CH_2(CH_2CH_2)_{n+1}CH_2CHPh_2 \xrightarrow[\text{2. film casting}]{\text{1. PE-H, 110 }^\circ C,\ PhCH_3} \text{PE-H/}CH_3CH_2(CH_2CH_2)_{n+1}CH_2CHPh_2$$

$$PE_{Olig}\text{-}CHPh_2 \qquad\qquad PE/PE_{Olig}\text{-}CHPh_2$$

(5)

$$\xrightarrow{BuLi} \text{PE-H/}CH_3CH_2(CH_2CH_2)_{n+1}CH_2CPh_2Li \xrightarrow{C_{60}} \text{PE-H/}CH_3CH_2(CH_2CH_2)_{n+1}CH_2CPh_2C_{60}$$

$$PE/PE_{Olig}\text{-}CPh_2Li \qquad\qquad PE/PE_{Olig}\text{-}CPh_2C_{60}$$

In the synthesis of an appropriately functionalized anionic macroinitiator, anionic oligomerization of ethylene[26] was followed by quenching of the living lithiated ethylene oligomer PE_{Olig}-Li with 1,1-diphenylethylene to form a diphenylmethyl terminated ethylene oligomer after protonation with methanol. About 75% of the oligomers so formed were functionalized with diphenylmethyl groups. The polyethylene degree of polymerization varied in different experiments but ranged between 1600 - 2500. Polyethylene films containing this oligomer were then formed by solution casting from a solution of PE_{Olig}-CHPh$_2$ and virgin polyethylene.[22] Surface grafting chemistry was effective with loadings of the functionalized oligomer to the host polyethylene that varied from 0.2 - 10 weight percent. Based on prior examples, we expected most of the diarylmethyl groups of the entrapped oligomers in these films to be at the polyethylene/solution interface and accessible to a soluble strong base like *n*-BuLi in THF.[9,10] Separate experiments showed that a soluble analog of PE/PE_{Olig}-CLiPh$_2$, 1-lithio-1,1-diphenylhexane prepared from the reaction of BuLi-TMEDA and 1,1-diphenylethene, successfully polymerized methacrylonitrile.[30]

Surface grafting of either PE/PE_{Olig}-CHPh$_2$ with methacrylonitrile proceeded by initial treatment with 0.04 N BuLi in THF. This produced a faint pink color on the film characteristic of 1,1-diphenyl-1-lithiohexane. Washing twice with fresh THF followed by addition of a 4 M THF solution of methacrylonitrile then yielded a surface grafted product.
Evidence for surface graft formation was obtained from ATR-IR spectroscopy, transmission IR spectroscopy, XPS spectroscopy and contact angle measurements. These data are summarized in Table 2. In the case of the anionic grafting, the amount of surface grafting increased with increasing reaction time. However, our initial efforts at continuing polymerization beyond 8 h have not been successful. We believe this reflects ad-

ventitious quenching of anionic sites. Analysis of the extent of surface grafting by transmission IR spectroscopy through measurement of the integrated absorbances of the nitrile -CN peak (2240 cm^{-1}) to a polyethylene - CH$_2$- peak (2020 cm^{-1}) and calculations using a Beer's law plot for known mixtures of poly-

Table 2. Anionic Grafting of Methacrylonitrile onto a
Polyethylene Film Containing Polyethylene-
Diphenylmethyl-terminated Ethylene Oligomer
Groups as Macroinitiators.

Time (h)	Monomer (M)	Degree of Polymerization	$C_{N_{1s}}/C_{Cl_{2p}}$	Contact Angle ($^\circ$)
0	4.0	0	0.0016	110
1	4.0	14	0.0332	92
2	4.0	21	0.0448	81
8	4.0	51	0.0921	71

ethylene and polymethacrylonitrile showed that the average degree of polymerization of the surface graft was ca. 50 after 8 h. This is likely a minimal value since some of the initiator sites in PE/PE$_{Olig}$-CHPh$_2$ probably are not effective participants in the surface polymerization chemistry. In any case, XPS spectroscopy suggests that the surface was largely polymethacrylonitrile. Further, contact angle analysis using water showed that the surface hydrophilicity (7_a 71°) was comparable to that of a pure cast polymethacrylonitrile film (θ_a 68°). A more direct assay of the degree of polymerization by dissolution and solution state analysis of the modified film was impractical because of the low concentration of modified polyethylene in a typical 1 cm^2 piece of film.
The absence of homopolymerization in either anionic or radical surface grafting was confirmed by ^{1}H NMR or IR spectroscopic analysis of the residue of the evaporated supernatant. In all cases, the surface grafted films were extracted with acetone in a jacketed Soxhlet apparatus for 6 h. These extractions and control experiments which showed that polymethacrylonitrile prepared in solution in the presence of an unfunctionalized polyethylene film using either *n*-butyllithium or 1-lithio-1,1-diphenylhexane was not adsorbed on polyethylene film established that the surface grafts were firmly attached to the polyethylene. Additional control experiments showed no surface modification occurred when unfunctionalized polyethylene films were treated with either *n*-BuLi and methacrylonitrile. The dispersity of the graft polymerization was not measurable in either the anionic or radical polymerization chemistry.

Preliminary work indicates that other monomers like acrylonitrile and acrylate esters can be substituted for methacrylonitrile in this chemistry. These studies and additional spectroscopic studies of the details of this surface grafting chemistry and of the surface grafted surfaces are ongoing.

Films containing a lithiated diphenylmethyl group were also useful in covalent attachment of C_{60} to polyethylene. This was accomplished by first preparing a lithiated film as outlined above. This lithiated film was then washed twice with THF to remove any residual *n*-BuLi and was then treated with 20 mL of a 4.6×10^{-4} M solution of C_{60} in heptane/THF (90:10) for 16 h with stirring to form $PE/PE_{Olig}-CPh_2C_{60}$ (equation 6). After quenching with CH_3OH, the product film was washed several times with hexanes and chloroform and was finally extracted for 4 h with hot chloroform using a Soxhlet apparatus to remove any absorbed C_{60}.

The X-ray photoelectron spectrum (XPS) of the film provided good evidence for incorporation of C_{60} at the polyethylene surface. This spectrum contained a lower energy shoulder on the normal carbon 1s peak. Curve fitting showed that this second peak was centered at ca. 285 eV. This is the reported position for a C_{60} peak in XPS spectroscopy[31] and differed from the major C_{1s} XPS peak which was a ca. 287 eV. This lower energy C_{1s} shoulder was not present in the starting film and its presence establishes that a there is a substantial C_{60} component in the outermost 50-100 angstroms of the film. In ongoing work, we are trying to use these covalently attached C_{60} groups as electrophiles to prepare a cross-linked grafted network at a polyethylene film surface.[32]

$$PE/PE_{Olig}-CHPh_2 \qquad PE/PE_{Olig}-CLiPh_2 \qquad PE/PE_{Olig}-CPh_2C_{60} \tag{6}$$

4   Acknowledgments

The work from our laboratory discussed in the above paper was made possible by support from the National Science Foundation and the Texas Advanced Research Program. It represents the efforts of a very competent and hardworking group of graduate students and postdoctorals whose names are explicitly noted in the cited references to our work.

5 References

1. W. J. Feast and H. S. Munro, Eds., "Polymer Surfaces and Interfaces", John Wiley and Sons: Chichester, 1987.
2. J. D. Swalen, D. L. Allara, J. D. Andrade, E. A. Chandross, S. Garoff, I. Israelachvili, T. J. McCarthy, R. Murray, R. F. Pease, J. F. Rabolt, K. J. Wynne and H. Yu, *Langmuir*, 1987, *3*, 932.
3. G. M. Whitesides, *Chimia*, 1990, *44*, 310-311.
4. D. E. Bergbreiter, in "Chemically Modified Surfaces", H. A. Mottola, J. R. Steinmetz, Eds., Plenum Press, New York, 1992, 133-154.
5. G. M. Whitesides and G. S. Ferguson, *Chemtracts: Org. Chem.*, 1988, *1*, 171.
6. R. R. Rye and G. W. Arnold, *Langmuir*, 1989, *5*, 1331 and references therein.
7. K. W. Lee, S. P. Kowalczyk and J. M. Shaw, *Macromolecules*, 1990, *23*, 2097.
8. R. R. Thomas, S. L. Buchwalter, L. P. Buchwalter and T. H. Chao, *Macromolecules*, 1992, *35*, 4559.
9. G. F. Xu, D. E. Bergbreiter and A. Letton, *Chem. Mat.*, 1992, *4*, 1240.
10. T. Corner, *Adv. Polym. Sci.*, 1984, *62*, 95-142.
11. W. K. Huang and G. H. Hsiue, *J. Polym. Sci., Polym. Chem. Ed.*, 1989, *27*, 3451-3463.
12. M. Suzuki, A. Kishida, H. Iwata and Y. Ikada, *Macromolecules*, 1986, *19*, 1804-1808.
13. R. H. Berg, K. Almdal, W. B. Pedersen, A. Holm, J. P. Tam and R. B. Merrifield, *J. Am. Chem. Soc.*, 1989, *111*, 8024-8026.
14. K. Allmer, J. Hilbron, P. H. Larsson, A. Hult and B. Ranby, *J. Polym. Sci., Polym. Chem. Ed.*, 1990, *28*, 173-183.
15. S. L. Regen, P. Kirszensztejn and A. Singh, *Macromolecules*, 1983, *26*, 335.
16. M. E. Galvin and G. E. Wnek, *J. Polym. Sci., Polym. Chem. Ed.*, 1983, *21*, 2727-37.
17. D. E. Bergbreiter and J. Zhou, *J. Polym. Sci., Polym. Chem. Ed.*, 1992, *30*, 2049-2053.
18. H. C. Brown, "Organic Syntheses Via Boranes", Wiley, New York, 1975.
19. The reaction of boranes with oxygen is an old reaction used originally in the late 1950's with triethylborane initiators. More recently organic synthesis applications of Et_3B have rediscovered this chemistry, cf. reference 20.
20. T. C. Chung and G. J. Jiang, *Macromolecules*, 1992, *25*, 4816.
21. D. E. Bergbreiter, C. Zapata and G. F. Xu, unpublished work.

22. D. E. Bergbreiter, Z. Chen and H. -P. Hu, Macromolecules, 1984, 17, 2111.
23. D. E. Bergbreiter, H. -P. Hu, M. D. Hein, Macromolecules, 1989, 22, 654-662.
24. D. E. Bergbreiter and M. D. Hein, Macromolecules, 1990, 23, 764-769.
25. D. E. Bergbreiter, M. D. Hein and K. J. Huang, Macromolecules, 1989, 22, 4648.
26. D. E. Bergbreiter, J. R. Blanton, D. R. Treadwell, R. Chandran, K. J. Huang, S. A. Walker and M. D. Hein, J. Polym. Sci., Polym. Chem. Ed., 1989, 27, 4205-4226.
27. D. E. Bergbreiter and B. Srinivas, Macromolecules, 1992, 25, 636-643.
28. D. E. Bergbreiter, B. Srinivas and H. N. Gray, Macromolecules, 1993, 26, in press.
29. C. H. Bamford, P. A. Crowe and R. P. Wayner, Proc. Royal Soc., A. 1965, 284, 455-468.
30. M. V. Beylan, S. Bywater, G. Smets, M. Szwarc and D. J. Worsfold, Adv. Polym. Sci., 1988, 86, 87-143. P. Rempp, E. Franta, J. -E. Herz, Ibid., 1988, 86, 145-73.
31. D. L. Lichtenberger, K. W. Nebesny and C. D. Ray, Chem. Phys. Lett., 1991, 176, 203.
32. F. Wudl, Acc. Chem. Res., 1992, 25, 157.

The Synthesis and Properties of Mutually Interpenetrating Organic–Inorganic Networks

Bruce M. Novak,* Mark W. Ellsworth, and Celine Verrier

DEPARTMENT OF CHEMISTRY, UNIVERSITY OF CALIFORNIA, BERKELEY, CA 94720, USA

1 INTRODUCTION

Only a small number of pure materials embody all of the physical and mechanical properties required for a given application. Because of this, most engineering and high technology materials are used either as blends, alloys, or composites, with the hope that the resulting compilation will retain the desirable characteristics of all of its constituents.[1,2] Under the right conditions, new or enhanced properties may emerge from these combinations of components. Furthermore, the properties of a composite material depend not only upon the properties of the individual components, but also upon the composite's phase morphology and interfacial properties. These interfacial interactions are foremost in determining the overall properties of the material. For example, the combination of a brittle fiber in a brittle matrix (i.e., glass fibers in an epoxy resin) to produce a superior material that is much tougher than either of the two single components, can be accomplished by the addition of coupling agents which mediate the interactions between the phases.[3] This synergism is governed by interfacial properties and is achieved by a combination of mechanisms that tend to keep cracks small and isolated, and which transfer and dissipate energy.

One approach to increasing interfacial interactions is to blur the ordinarily sharp interfacial zone by synthesizing materials which show a high degree of mixture, or interpenetration, between the two dissimilar phases. To this end, we, as well as others, have been interested in using the sol-gel process[4,5] to develop new routes into hybrid materials which contain both inorganic and organic components commingled into intimate, new morphologies. Phase domain sizes are minimized through these processes (sometime approaching the molecular level), and highly transparent composite materials often result.

Historically, the sol-gel process involves an acid or base catalyzed hydrolysis and condensation of tetraalkoxy orthosilicates ($Si(OR)_4$) to yield solvent swollen SiO_2 gels (equation 1).

$$Si(OR)_4 \xrightarrow{\ H_2O/H^+\ } \text{[Si–O–Si network]} \equiv \begin{array}{l} \text{Solvent-Swollen} \\ SiO_2 \text{ Matrix} \\ \bullet \text{ Excess Water} \\ \bullet \text{ Alcohol} \\ \bullet \text{ Cosolvent} \end{array} \quad (1)$$

Ultra-slow drying of these gels under controlled, ambient conditions leads to crack-free, monolithic glass formation. One of the major obstacles to the wide-spread application of sol-gel techniques is the fact that this drying process is accompanied by extraordinary shrinkage of the glass (shrinkages of > 75% are common). We will herein describe our work on the simultaneous formation an organic polymer within these sol-gel derived, inorganic networks (simultaneous interpenetrating networks) in order to substantially increase their strength and eliminate the large scale shrinkages associated with solvent removal during the drying process.

RESULTS AND DISCUSSION

Three recent publications[6,7,8] describe our approach to the formation of unique hybrid materials. The basic approach involves the formation of inorganic networks (SiO_2) through a sol-gel process using tetraalkoxysilane with polymerizable alkoxide moieties. By employing *in situ* organic polymerization catalysts (ROMP or free radical), the alcohol liberated during the formation of the inorganic network is polymerized. By using a stoichiometric quantity of water, and additional polymerizable alcohol as a cosolvent if needed, all components are converted into either the organic polymer or the inorganic network. Because no evaporation is necessary, large scale shrinkages are eliminated (Scheme I).

Scheme I

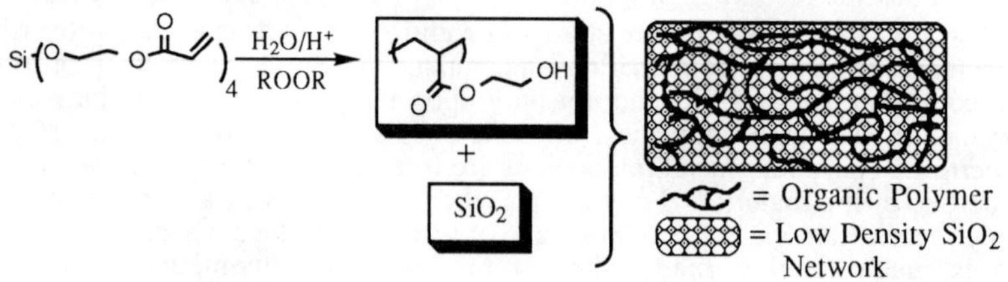

= Organic Polymer
= Low Density SiO_2 Network

The glass content of these composites is controlled by the stoichiometry of the tetraalkoxysilane precursor and typically ranges from 10-15%. Current work in this area is directed at increasing the glass content in these composites by synthesizing poly(silicic acid ester) derivatives possessing the same polymerizable alkoxides.[8] By controlling both the silicic acid branching ratio and the degree of alkoxide substitution, nonshrinking composites with glass contents greater than 50% can be fabricated (Figure 1).

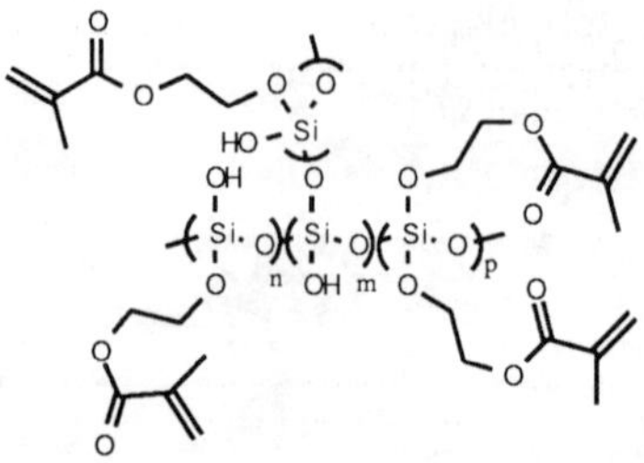

Figure 1. Poly(silicic acid) ester possessing polymerizable alkoxides.

These simultaneous interpenetrating network (SIPN) composites display excellent properties in comparison to the pure polymer. For example, intertwining the inorganic network into 2-hydroxyethyl methacrylate (HEMA) at a level of 27 volume percent increases the yield strength by over 835% (30 MPa vs. 281 MPa), and the elastic modulus by over 140% (900 MPa vs. 2.2 GPa). Furthermore, the mechanical properties of these materials compare very favorably with those of "conventional" glass/polymer composites. A direct comparison of these SIPNs to a chopped fiber-HEMA can be seen in Figure 2.

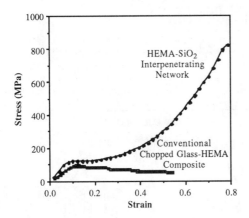

Figure 2. Comparison of the compressional strength of HEMA-SiO_2 interpenetrated network composite and a conventional chopped glass fiber-HEMA composite.

Mechanical testing on these materials indicates that they behave neither as polymers nor as ceramics. Compared to ceramic glasses, these composites at comparable stress values typically show 10^3-fold increases in toughness (2.6 X 10^3 increase for 50% glass). Again, these composites do not obey the "rule of mixtures", but rather show synergistic nonlinear behavior with respect to toughness (Figure 3).

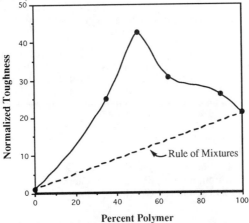

Figure 3: Plot of the normalized toughness of a series of HEMA-SiO_2 interpenetrated network composites.

The mechanical properties of these materials depends upon both the percent glass as well as the nature of the organic polymer. An illustration of the importance of this latter variable can be seen by comparing the properties of composites prepared using HEMA ($T_g \approx 110$ °C) with composite prepared using 2-hydroxyethyl acrylate (HEA) ($T_g \approx -15$ °C).

HEMA

Tg = + 110 °C

HEA

Tg = - 15 °C

As can be seen in Figure 4, these materials behave as either stiff, high modulus materials (HEMA) or elastic rubbers (HEA).

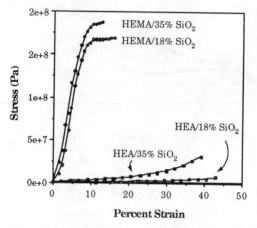

Figure 4. The stress-strain curves (elastic regime) for interpenetrating network composites prepared with HEMA and with HEA.

CONCLUSIONS

The convenient and mild conditions offered by the sol-gel process for the preparation of inorganic networks has paved the way for the synthesis of new materials that display unique properties. The versatility offered by synthesizing organic polymers within these networks allows for the synthesis of transparent composites with mechanical properties that range from rubbery elastomers to high modulus materials.

ACKNOWLEDGMENTS

The author gratefully acknowledges financial support for this work from the Office of Naval Research, the NSF Presidential Faculty Fellowship, the Alfred E. Sloan Foundation, Du Pont, BFGoodrich, Dow Corning, and the Corning Corporation.

REFERENCES

1. J. A. Manson, L. H. Sperling, "Polymer Blends and Composites," Plenum Press, New York, 1976.
2. G. Lubin (Ed.): "Handbook of Composites", Van Nostrand Reinhold Co., New York, 1982, Volumes 1 and 2.
3. E. P. Plueddemann, "Silane Coupling Agents," 2d Edition, Plenum Press, New York, 1991.
4. C. J. Brinker, G. W. Scherer: "Sol-Gel Science," Academic Press, New York, 1990.
5. D. R. Ulrich, Chemtech, 1988, 242.
6. Ellsworth, M. W.; Novak, B. M. J. Am. Chem. Soc. , 1991, 113, 2756.
7. Novak, B. M.; Davies, C. Macromolecules , 1991, 24, 5481.
8. Ellsworth, M. W.; Novak, B. M. Chem. Mater., 1993, 5, 839.

The Physisorption and Condensation of Aminosilanes on Silica Gel

Karl C. Vrancken,* P. Van Der Voort, K. Possemiers,
P. Grobet,[1] and E.F. Vansant
LABORATORY OF INORGANIC CHEMISTRY, UNIVERSITY OF ANTWERP
(UIA), UNIVERSITEITSPLEIN I, B-2610 WILRIJK, BELGIUM
[1] LABORATORY OF SURFACE CHEMISTRY & CATALYSIS, CATHOLIC
UNIVERSITY OF LEUVEN, CARD. MERCIERLAAN 92, B-3001
HEVERLEE, BELGIUM

1 ABSTRACT

The modification of silica gel with aminosilanes has been studied by UV spectrometry and Cross Polarisation Magic Angle Spinning Nuclear Magnetic Resonance (CP MAS NMR) and Fourier Transform Infrared Photoacoustic (FTIR-PAS) spectroscopy. A quantitative determination of the amount of physisorbed molecules and evaluation of the coating stability led to a further insight in the reaction mechanism of γ-aminopropyltriethoxysilane (APTS) and γ-aminopropyldiethoxymethylsilane (APDMS) with silica gel. The availability of surface hydroxyls was evaluated in the modification of deuterated silica gel. Intraglobular silanols can not be reached by the aminosilane molecules. In the final coating the majority of the aminosilane molecules are bonded with two siloxane bonds to the silica gel surface and have the amino group in a free form.

2 INTRODUCTION

Aminoorganosilanes (H_2N-R-Si(OR')$_3$) play an important role in the surface modification of oxide materials. The most widely used and studied aminosilane is γ-aminopropyltriethoxysilane (APTS, $H_2NCH_2CH_2CH_2Si(OCH_2CH_3)_3$). Its reaction with various types of silica has been the topic of much research. Valuable information on the role of the amine group and the alkoxy functions can be obtained by comparing the reaction behaviour of APTS to n-butylamine and γ-aminopropyldiethoxymethylsilane (APDMS, $H_2NCH_2CH_2CH_2Si(CH_3)(OCH_2-CH_3)_2$).

Compared to other organosilanes, aminosilanes show an increased reactivity owing to the presence of the nucleofilic amine functionality. The amine group causes an instant hydrogen bonding of the silane molecule to the oxide material.[1] After this physisorption, the amine can catalyse the direct condensation of the silane with the surface hydroxyl groups, with loss of an alcohol.[2] In the presence of water, however, the alkoxy groups will quickly hydrolyse to silanols.[3] Hydrolysed silane molecules condense to form oligomers, or react with the solid substrate to form siloxane bonds. When water is used as a solvent, this condensation will lead to a polymerised coating of uncontrollable thickness.

When the reaction is performed in a dry organic solvent, the condensation reactions in the reaction step can be minimized. Aminosilane molecules are merely physically bonded to the substrate surface. Therefore, the coated substrate is submitted to a thermal curing after reaction, driving the chemisorption, i.e. formation of siloxane bonds, to completion. This conversion has been established by various authors.[4,5] Curing atmosphere and temperature play an important role. Vandenberg et al.[6] used various techniques to identify the micro- and macrostructure changes upon curing in various atmospheres. They succeeded in a qualitative description of the modification of thin silicon oxide layers with APTS. NMR studies have been performed to study the parameters governing the curing process. From ^{13}C NMR peak shift positions, Chiang et al.[7] concluded that heat treatment destroys 60% of the hydrogen bonding of APTS, deposited from aqueous solvent. Kelly and Leyden[8] studied the amine interaction with the silica substrate by thermometric enthalpy titration and were able to distinguish hydrogen bonding interaction from ionic proton transfer.

Insight in the conversion of physisorberd silane to chemisorbed species can be obtained by studying the stability of the coating. Upon stirring of the coated substrate in ethanol, physisorbed silane molecules will desorb with a different rate than chemisorbed ones. The rate of desorption can be determined with a color reaction of the amine group with salicylic aldehyde. This method was proposed by Wadell et al.[9] and led to the proposal of various bonding models for air cured modified silica. Despite of the many studies on this modification, some topics in the reaction mechanism remain unresolved. This is due to the large number of reaction parameters involved

A combination of NMR, FTIR and UV spectrometry enables to study and quantify the modification of silica gel with APTS and APDMS in terms of physisorption and chemisorption.

3 EXPERIMENTAL

1g of dehydrated (673K, 20h) Kieselgel 60 (Merck) was stirred for 2h with 50ml of a 1%v/v APTS (A1100, UCAR) or APDMS (Hüls AG) solution in toluene. For n-butylamine (Fluka) modification an equimolar 0.4%v/v solution was used. The reacted substrate was filtered and dried for 5min. at 383K. Curing was performed for various times in air at 383K or under vacuum at 423K. For UV tests 1.000g of modified silica was stirred in 50ml 0.3%salicylic aldehyde/ethanol (Merck) solution. At indicated times a 5ml sample was taken, centrifuged and the supernatant was measured at 404nm. For deuteration, D_2O vapor was pumped through the dehydrated silica sample, in a dynamic adsorption apparatus, at 673K for 1h. The sample was subsequently evacuated at room temperature for 30min. The [29]Si (79.5MHz) and the [13]C (100.6MHz) CP MAS NMR were performed on a Bruker 400 MSL spectrometer. The [29]Si CP experiments were performed using a contact time of 5ms, a recycle time of 3s, a spinning rate of 3.5kHz and a number of scans between 3000 and 8000. In the ^{13}C CP MAS experiment the contact time was 2.5ms, the recycling time 3s, the spinning rate 4kHz and the number of scans 10000. The FTIR-PAS measurements were performed, using a prototype MTEC-100 photoacoustics cell (McClelland). Samples were loaded under dry nitrogen atmosphere. Spectra were recorded using a 20SX Nicolet FTIR spectrometer at 4cm^{-1} resolution. Carbon black was used as a background.

4 RESULTS AND DISCUSSION

The modification of silica with aminosilanes proceeds in three steps. (i) A thermal pretreatment of the silica gel determines the degree of hydration and hydroxylation. (ii) In the reaction step the aminosilane is stirred with the silica in an appropriate solvent. (iii) The modified substrate is treated thermally to stabilise the coating (curing). In this last step, the presence of air humidity is an important parameter. Therefore clear distinction will be made whether samples have been cured on air or under vacuum.

UV spectrometric stability study of non-cured and cured modified silicas.

In the reaction phase of the modification of silica gel with APTS in dry solvent a monolayer coating is formed.[10] The monolayer capacity has been determined as 1.0mmol/g dry silica in a previous publication. Within the concentration limits used, the monolayer formation involves a homogeneous distribution of the silane molecules over the silica surface, irrespective of the presence of functional groups. The amine function is responsible for the fast adsorption of the silane molecules on the silica surface.

In ethanol solvent, amines react with salicylic aldehyde with the formation of a yellow Schiff's base, having $\lambda_{max} = 404$nm. After calibration, the amine function thus allows the quantitative determination of the amount of an alkylamine or aminosilane liberated from a modified substrate upon ethanol leaching, by UV detection. The role of the amine function in the stability of the coating is studied by comparing the behaviour of silica modified with n-butylamine to APTS modified silica. n-Butylamine has no other interaction possibilities with the silica silanols but the amine group. From Kjeldahl analysis a surface coverage of 1.2mmol/g of butylamine on silica was determined.

In Figure 1 the amount of n-butylamine and APTS liberated from 1.00g of air dried modified silica is compared. The amount of product lost as well as the course of the loss, upon stirring of the modified substrate in ethanol, are indicative of the stability of the coating. Both samples show a rapid initial loss. For APTS (Fig. 1) this is followed by a slow process.

The APTS appears to have a largely increased stability compared to the butylamine. A tenfold amount of butylamine is released from the silica, over the measured time interval. Starting with a coverage of 1.15 mmol of butylamine and 0.82mmol of APTS per g of modified silica, a relative amount of respectively 78% and 11% is lost after 6 days of stirring in ethanol.

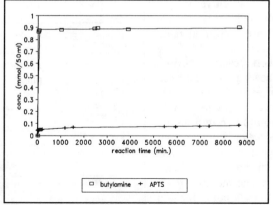

n-Butylamine can interact with the silica surface hydroxyls by hydrogen bonding as well as proton transfer. The hydrogen bonded species are easily removed by

Figure 1: loss curves for butylamine and APTS upon prolonged ethanol leaching

ethanol. 22% of the coating is stable towards ethanol leaching and therefore is concluded to be bonded by ionic interactions.

An identical chemical behaviour of the amine group of butylamine and APTS may be assumed. Therefore ionic interactions will have an equal share (22%) in the surface bonded APTS molecules. Kelly and Leyden[8] studied the interaction of APTS with silica gel by thermometric enthalpy titration. They found that 26% of the APTS molecules were irreversibily bonded to the silica surface and attributed this stability to ionic interactions. Both values are equal within experimental error.

The increased stability of APTS compared to butylamine is attributable to interactions involving the silicon side of the molecule. Most of the APTS molecules are not only hydrogen bonded by the amine function. Besides this weak interaction, other firmer interactions appear to play an important role already before curing. Care should be taken however, because this test concerns an air dried substrate. Reactive adsorbed species may rapidly condense or hydrolyse upon exposure to air humidity. Therefore the coating stability might be slightly overestimated. The drying step can however not be overcome.

Before curing a monolayer of APTS resides on the silica surface. 22% of the molecules are bonded by ionic interaction of the amine group. At least 10% of the coating is bonded by hydrogen bridge formation with silica silanols. Considerable condensation already occurs in the reaction phase.

In order to study the condensation process in the curing phase, more thoroughly, experiments at short ethanol leaching times were set up. The loss curves of APTS for variably cured samples are displayed in Figure 2. Again, the position and profile of the absorbance curves are indicative for the stability of the coating under study.

The curve of the uncured sample (x) shows a clear increase over this relatively short reaction period, showing the occurrence of an easily removable and a more firmly bound component. However both are not stable towards short term ethanol leaching. For the sample which has been cured for 20h under vacuum (□), thus securing the absence of air humidity, the coating is stabilised. An invariable amount of silane is detected over the entire leaching period, indicating

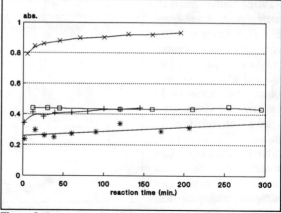

Figure 2: loss curves for short term ethanol leaching on APTS modified silica: (x) uncured; (+) 3h air cured; (*) 20h air cured; (□) 20h vacuum cured.

only one type of physical interaction. The increased coating stability after 20h of curing is due to chemisorption to the silica surface. Still a small amount of the coating molecules is easily removed. The absorbance value found corresponds to a concentration of 0.02mmol/g. In other words only 2% of the coating is not irreversibly bound to the surface. These molecules are physisorbed to the surface by hydrogen bonding of the amine group to a surface hydroxyl. For the sample cured in air for 20h (*), an additional stabilisation is obtained. This stabilisation is attributed to oligomerisation of the silane molecules on the silica surface. Oligomerisation occurs under the influence of air humidity. From the difference in position of the curves for 3h (+) and 20h (*) cured sample, it can be concluded that there is a progressive oligomerisation during the air curing period.

The air cured APTS curves have a more clear sloping. In the air cured APTS the rapid release of loosely physisorbed molecules is followed by a slower loss. Because the slow process is measurable within this short ethanol leaching time, it can not be attributed to the breaking of chemical bonds. The occurrence in humid conditions led us to conclude that this slow loss is caused by hydrogen bonding interactions of the silane molecules via the silane silanols. These silanols are formed under the influence of air humidity.

Concluding it may be stated that the amount of non-chemically bonded silanes decreases upon curing of the samples, following condensation with the surface. Extra stabilisation is obtained by oligomerisation of the silane molecules, for air cured samples. Also for these samples, silane hydroxyls, formed upon hydrolysis of ethoxy groups, are in hydrogen bonding interaction with the silica surface. Figure 3 gives a clear view on the stabilisation of the coating by condensation with the surface hydroxyls. The amount of hydrogen bonded aminosilane, as calculated from the absorbance data, is plotted as a function of curing time in vacuum.

In evaluating the data for the vacuum c u r i n g   o n l y , contributions from oligomerisaion and hydrogen bonding of silane silanols are ruled out. A d d i t i o n a l information is obtained in comparing the data for APTS modified silica to those measured on silica modified with APDMS. The APDMS molecule has only two ethoxy groups and therefore can only form two chemical bonds to the silica surface.

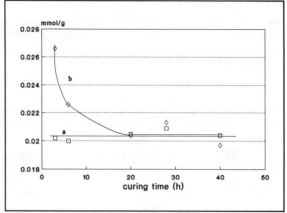

Figure 3: amount of physisorbed silane molecules per g of modified silica as a function of curing time in vacuum, (a) APTS; (b) APDMS.

For APTS a maximal stability (minimal amount of physisorbed species) is obtained within 3h of curing at 150°C under vacuum. In the APDMS case the physisorbed amount decreases with curing time increasing up to 20h. Because of the presence of only 2 ethoxy groups in the APDMS molecule, compared to 3 for the APTS, stabilisation by formation of a siloxane bond is at maximum only after prolonged curing. For vacuum curing times above 20h, the same level as obtained for APTS is maintained.

An amount of 0.02mmol of aminosilane remains merely hydrogen bonded even at very long curing times in vacuum. This is however only 2% of the totally deposited amount.

NMR and FTIR-PAS study of the chemically bonded species.

After characterisation of the physisorption, we wish to get information on the chemical structure of the coating. ^{29}Si and ^{13}C CP MAS NMR spectra of 20h air and vacuum cured APTS modified silica and 20h air cured APDMS modified silica were recorded.

The ^{29}Si NMR spectra (Fig.4) give information on the coordination at the silicon of the silane molecules and in the silica gel. Band deconvolution is used to obtain the peaks as reported in Table 1.

In the vacuum cured sample (Fig.4a), the APTS is mainly in the bidentate form (-59ppm). From curve fitting and integration a relative amount of 60% was calculated. Monodentate and tridentate forms show only minor contributions (20% each). The previously discussed hydrolysis and oligomerisation are evident from the spectrum of the air cured sample (Fig.4b). The silane region shows clear contributions from the hydrolysed monodentate (-46ppm) and oligomerised (tridentate, -66ppm) forms. The APDMS is found to be almost entirely (90%) in the bidentate form. Because of the bifunctionality of the APDMS this is the most plausible form. Any hydrolysis and condensation between silane molecules can at most lead to surface dimers. A minor contribution of the monodentate form (-12.4ppm) is visible.

In the high ppm region of the ^{29}Si spectra, the silica Si atoms are visible. Signals at -93 and -100 ppm indicate that some silica silanols are left after the entire modification procedure. An explanation for this is found in the ^{13}C NMR and FTIR data, discussed below.

The ^{13}C NMR spectra give information on the presence of the ethoxy groups and on the mobility of the aminopropyl chain. Peak positions and assignments are given in Table 2.

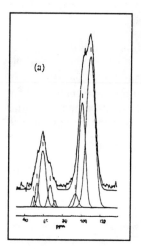 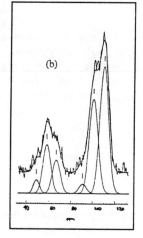 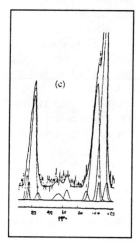

Figure 4: ^{29}Si CP MAS NMR spectra of modified silica: (a) APTS modified, vacuum cured; (b) APTS modified air cured; (c) APDMS modified, air cured

The vacuum cured APTS sample (Fig.5a) has major ethoxy peaks. The coating molecules clearly have ethoxy groups left. Air curing (Fig.5b) causes a near complete loss of ethoxy groups, because of oligomerisation, as stated before. In the air cured APDMS sample (Fig.5c) no ethoxy character is left at all.

Table 1: ^{29}Si CP MAS NMR peak positions (ppm) and assignments. (R = CH$_2$CH$_2$CH$_2$NH$_2$; R'=H or CH$_2$CH$_3$)

silane used + curing conditions	chemical shift (ppm)						
APTS vac 20h	(-49)	-53	-59	-67	-93	-100	-103
APTS air 20h	-46		-58	-66	-90	-100	-110
assignment					geminal OH	lone OH	silox-ane
APDMS air 20h	-19.7				-90	-100	-110
assignment							

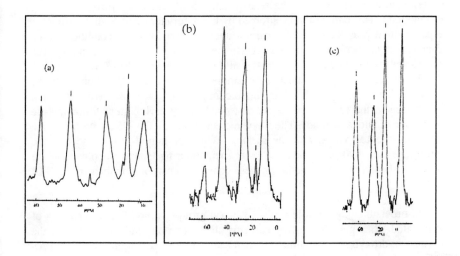

Figure 5: ^{13}C CP MAS NMR spectra of modified silica: (a) APTS modified, vacuum cured; (b) APTS modified air cured; (c) APDMS modified, air cured

Table 2: ^{13}C CP MAS NMR peak positions (ppm) and assignments

silane used + curing conditions	chemical shift (ppm)					
APTS vac 20h	57.8	43.5	26.6	15.8	8.4	
APTS air 20h	(57.4)	42.7	25.0	(15.5)	8.0	
APDMS air 20h		43.2	24.4		12.9	-4.6
assignment	-OCH$_2$-	γ	β	-CH$_2$-CH$_3$	α	Si-CH$_3$

The peak position of the β-carbon is affected by the mobility of the aminopropyl chain. Some ambiguity has been found in literature on this topic.[7,11,12] The type of interaction of the NH$_2$ group clearly determines the β-carbon peak position. It has been clearly established that protonation of the amine by a surface hydroxyl causes a shift of the peak to 20-21ppm. This protonation is favoured in the presence of water. The peak at 25-27ppm has been assigned to free as well as hydrogen bonded amine. The present data indicate a shift to lower ppm value in the 25-27 region, for the sample cured in vacuum compared to the air cured sample. The β-carbon of a non-cured sample was found at 24.7ppm. It can therefore be concluded that the position of the β carbon peak is indicative for the hydrogen bonding interaction of the amine function. The peak shift is determined by a variation in the relative amount of free and hydrogen bonding interactions. This conclusion is enforced by the observations of Sudhölter *et al.*[12] Apart from a peak shift of the β carbon under various washing conditions, they found a small peak at 29ppm in a sample in which hydrogen bonding was obstructed by the presence of trimethylsilyl species on the silica surface.

In the uncured sample most of the amine groups are hydrogen bonded to the surface. The hydrogen bonding interaction, which was needed for quick adsorption in the reaction phase, is destroyed in the curing step, following the formation of siloxane bonds. The silane molecules are on the silica with the amine function oriented away from the surface. When air humidity is present in the curing phase, hydrogen bonding with neighbouring silane silanols can be accomplished. Further grounds for these conclusion can be obtained from FTIR data.

In order to study the role of the hydroxyl groups, a modification reaction of deuterated silica gel was performed. This allows to study the availability of the surface hydroxyls for reaction with D$_2$O and APTS. In Figure 6, FTIR-PAS spectra of silica and modified silica are displayed.

In the spectrum of pure silica (Fig.6a) bands corresponding to free (3740cm^{-1}) and bridged (3650cm^{-1}) silanols are visible. Upon reaction with D$_2$O vapor (Fig.6b) all the free silanols and part of the bridged silanols are exchanged to

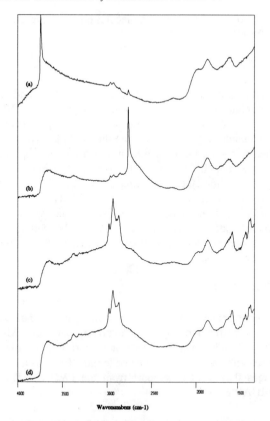

Figure 6: FTIR-PAS spectra : (a): dehydrated silica gel; (b): (a) after D₂O exchange; (c): (b) after reaction with APTS; (d): (a) after reaction with APTS.

≡Si-OD (2760 and 2730-2450cm^{-1} resp.). Any change in the deuteration parameters did not increase the exchange efficiency. The non-exchangeable bridged silanols are concluded to be unavailable for the D₂O vapor. They are residing inside closed cavities in the silica gel structure and are named intraglobular OH's.[13] Modification (Fig.6c) of the deuterated substrate leads to a complete removal of the free ≡Si-OD groups. Comparing the bridged ≡Si-OD region (2770-2630 cm^{-1}) to the spectrum of a non-deuterated APTS modified silica gel (Fig.6d), a very weak shoulder may be discerned. This can be attributed to a small percentage of the surface hydroxyls in hydrogen bonding interaction with an aminosilane amine group. Amine stretching and bending vibration bands are at 3376, 3311, and 1597 cm^{-1}. These positions are only slightly shifted, compared to the pure APTS spectrum (3380, 3315, 1603 cm^{-1}), indicating only a small contribution of hydrogen bonded species. For the vacuum cured sample the amine is almost completely in the free form.

Residual bridged surface silanols are present in the modified silica spectrum. Apparently, the bridged silanols which are not available for deuterium exchange can not be reached by the aminosilane molecules either. It are these silanols which appear in the ^{29}Si NMR spectrum also.

5 CONCLUSION

The modification of dehydrated silica gel with APTS and APDMS in dry solvent is fully characterized by the combination of UV, NMR and FTIR measurements. In the reaction phase a monolayer coating is formed on the silica surface. The aminosilanes can not reach the intraglobular hydroxyls. 22% of the monolayer molecules are bonded by proton transfer from a surface silanol to the amine group. More than 10% is bonded by hydrogen bonding of the amine group. Considerable condensation takes place already in the reaction step. Curing causes further condensation with the surface. Air humidity causes hydrolysis and oligomerisation. Silane silanols also form hydrogen bonds with the surface. When the sample is cured under vacuum, the molecules are bonded in a bidentate form, with the amine group pointed away from the silica surface. 2% of the coating remains merely physisorbed.

ACKNOWLEDGEMENT

L. De Coster is acknowledged for optimising the deuteration procedure and performing the modification of the deuterated samples. Mrs.H.Geerts of K.U.Leuven is gratefully acknowledged for performing and discussing the CP MAS NMR measurements. KCV is indebted to the NFWO as a research assistant. KP acknowledges the financial support from the IWONL. This research was funded by the IUAP.

REFERENCES

1. Morrall S.W., Leyden D.E., in Silanes, Surfaces and Interfaces, D.E.Leyden ed.,Gordon and Breach, New York, p.501.

2. Blitz J.P., Murthy R.S.S., Leyden D.E., J.Coll.Interfac.Sci., **126**(2),387,(1988).

3. Vrancken K.C., Van Der Voort P., Vansant E.F., Grobet P., J.C.S.Faraday Trans, **88**(21),3197,(1992).

4. De Haan J.W., Van Den Bogaert H.M., Ponjée J.J., Van De Ven L.J.M., J.Coll.Interfac.Sci., **110**,591,(1986).

5. Caravajal G.S., Leyden D.E., Maciel G.E., in Silanes, Surfaces and Interfaces, D.E.Leyden ed., Gordon and Breach, New York, 283,(1985).

6. Vandenberg E.T., Bertilsson L., Liedberg B., Uvdal K., Erlandsson R., Elwing H., Lundström I., J.Coll.Interfac.Sci., **147**,103,(1991).

7. Chiang C., Liu N., Koenig J.L., J.Coll.Interfac.Sci., **86**,26,(1982).

8. Kelly D.J., Leyden D.E., J.Coll.Interfac.Sci., **147**,213,(1991).

9. Wadell T.G., Leyden D.E., DeBello M.T., J.Am.Chem.Soc., **103**,5303,(1981).

10. Vrancken K.C., Casteleyn E., Possemiers K., Van Der Voort P., Vansant E.F., JCS Faraday Trans, **89**(12), (1993) in print.

11. Caravajal G.S., Leyden D.E., Quinting G.R., Macial G.E., Anal.Chem., **60**,1776,(1988).

12. Sudhölter E.J.R., Huis R., Hays G.R., Alma N.C.M., J.Coll.Interface Sci., **103**,554,(1985).

13. Iler R.K., The Chemistry of Silica, J.Wiley &sons, New York, (1979).

Extraordinary Thermal Stabilization of Enzymes through Surface Attachment by Covalently Bound Phospholipids

K. M. R. Kallury and M. Thompson
DEPARTMENT OF CHEMISTRY, UNIVERSITY OF TORONTO, TORONTO, ONTARIO M5S 1A1, CANADA

INTRODUCTION

The persistent demand for improved solid phase systems for applications in diversified areas has provided substantial impetus to research on the generation and characterization of chemically modified surfaces.[1-8] The need for surface reconstruction stems from the fact that unmodified surfaces present a number of problems detrimental to their intended use. A classical example is the development of different high performance liquid chromatographic stationary phases through the surface modification of silica, the surface hydroxyls of which interfere with the chromatographic performance of this naturally abundant material in the unmodified form.[9] Similarly, the inertness of the polymers polyethylene or polytetrafluoroethylene limits their full exploitation in the electronics and other engineering areas due to the basic problem of their adhesion to metals, making it necessary to render the surfaces of these polymers more hydrophilic in order to enhance their adhesive properties.[10,11] By the same token, implants, grafts and contact lenses made of teflon or other polymers suffer from the problem of rejection by the human system due to biological phenomena such as thrombogenesis or protein absorption, when contacted by blood or tissue, and hence there is a compelling need for the restructuring of these polymeric surfaces to improve their biocompatibility.[12,13] On the other hand, metals, ceramics or even wood products have to be rendered hydrophobic to combat corrosion and/or decay due to moisture.[14-16] Furthermore, the surfaces of materials used in nuclear reactors or space crafts also require modification in order to enhance their capability to withstand extreme conditions such as high temperature and pressure or radiation damage.[17-19]

In addition to the above areas of applicability, chemically modified materials also find wide utility as solid phase analytical reagents for clinical diagnosis/treatment[20] (biosensory applications), for environmental monitoring[21] (chemical sensors) and for the extraction/detection of metals/metal ions.[22] An appropriate molecule with selective binding properties is chemically bound to a suitable solid support for such purposes. Other fields where modified solid matrices are gaining attention include the anchoring of soluble catalysts (e.g. phosphorus and nitrogen-based ligands) on metal-complexed silica particles,[23] and controlled drug delivery systems which use liposomes or other biodegradable polymers.[24]

One major area of applicability of chemically modified surfaces lies in the immobilization of biomolecules (e.g. enzymes) on semiconductor or metallic or polymeric surfaces. The advantages of enzymes over synthetic catalysts are manifold. Enzymes possess high catalytic activity, substrate specificity, work under mild conditions with minimal by-product formation and pose little threat to the environment. Moreover, they can be produced in large amounts and in a majority of cases, are inexpensive. Due to all these factors, enzymes are becoming increasingly popular as biocatalysts in a variety of industries such as pharmaceuticals, food products and beverages.[25] Other important areas of utility include chiral organic synthesis,[26] waste treatment,[27] clinical diagnosis,[28] detection of toxic chemicals[29] and fundamental biochemical studies.[30] The practical use of enzymes often requires elevated temperatures to increase productivity, prevent microbial contamination and improve the solubility of substrates. Unfortunately, enzymes in their native state are unstable to elevated temperatures, as well as to aqueous-organic environments and to pH variations and lose most of their activity when subjected to such conditions either collectively or individually.[31] Various methods have been developed to combat this denaturation problem. These consist of functional group modifications, addition of stabilizing agents, use of organic solvents, genetic/protein engineering, immobilization to solid supports and isolation of enzymes from thermophilic organisms.[32] Of these, the most thoroughly investigated and proven to be the most successful technique is the immobilization of enzymes to solid supports.

Immobilization of enzymes can be achieved through covalent binding to functionalized solid support materials, by adsorptive interactions, entrapment into gels or beads or fibres, cross-linking or co-cross-linking with bifunctional reagents and encapsulation in microcapsules or membranes. Of these, the covalent attachment protocol has been demonstrated to be the most practicable and considerably enhances the thermal stability and shelf life of enzymes. As pointed out by Taylor,[33] although the art and science of immobilizing proteins in general and enzymes in particular is now the oldest of the new biotechnologies, it forms the basis for newly emerging biotechological products/processes such as immobilized antibodies, solid-phase synthesis of DNA, development of DNA hybridization probes and bioaffinity chromatographic reagents, to mention a few. An up-to-date account of various immobilization protocols has been presented by Cabral and Kennedy.[34]

Obviously, the immobilization of biomolecules to solid supports requires the chemical modification or functionalization of the latter if they do not inherently carry the necessary functionalities. Examples of carriers which do contain such functionalities are graphite, carboxymethylcellulose, poly(hydroxyethyl methacrylate) or chitosan. Functional groups most amenable to protein immobilization are the primary hydroxyls, thiols, carboxylic moieties and amino groups. Functionalities on the proteins themselves useful for immobilization purposes are the amino groups on lysine residues, the carboxyls on aspartic or glutamic acid residues, the thiols on cysteine residues and the hydroxyls on tyrosine or serine residues.

Solid supports could be modified to generate the requisite functionalities by protocols such as silanization,[35] treatment with functionalized thiols,[36] derivatization of electrochemically deposited polymers[37] (e.g. polypyrrole) or through physical techniques such as plasma treatment, corona discharge or metallation.[40] Silanization is the most extensively used method for introducing reactive moieties on to solid supports. For this purpose, three types of reagents are in vogue, viz. chlorosilanes, alkoxysilanes and silicon hydrides. Amongst these, the alkoxysilanes are the most stable and the easiest to handle, although they are less reactive than the other two classes of silanizing agents. The alkoxy moieties of silanes such as aminopropyltriethoxysilane (APTES) are stable to mild reaction conditions and thus a desired functionality could be inserted readily into an alkoxy silane. Such pre-derivatization procedures are not practicable with chloro- or hydrido-silanes and hence a surface reaction must preceed any attempted functionalizations with these two classes of silanes.

Although a number of published reports[41] have indicated that a phospholipid environment exerts considerable stabilizing influence on the enzymatic activity, no attempt has so far been made to covalently immobilize enzymes to such lipid matrices. Hence, we undertook an investigation on the thermal and storage stability of matrices on which the enzyme urease was covalently deposited on to phospholipid molecules which are already covalently attached to solid supports. For comparison, the same enzyme was also covalently bound to the same carrier surfaces in the absence of the lipid. The surfaces were all characterized by X-ray photoelectron spectroscopy (XPS), ellipsometry and FTIR. The thermal and storage stability of the enzyme was assessed in each case by UV spectrophotometry using an urea-bromocresol purple substrate solution. Silica, tungsten and teflon were chosen as representative solid supports.

RESULTS AND DISCUSSION

Our strategy to covalently immobilize urease on to silica, tungsten or teflon supports consists of initially derivatizing these support materials with an alkoxysilane which carries a carboxy or amino substituent at the terminus of a long alkyl or acyl chain attached to the silane silicon atom. This ω-substituent on the chain must be protected in order to avoid undesirable side reactions during the preparation of the functionalized silane. In order to effect the silanization, the support materials must carry hydroxylic functionalities. Silica inherently carries such surface hydroxylic groups, but tungsten and teflon do not. Hence, the tungsten support (1 sq.cm. thin strip) was oxidized thermally and then treated with sodium hydroxide to hydrate the surface oxide film, according to an earlier reported procedure.[42] Teflon was hydroxylated by a metal deposition technique developed in our laboratories earlier.[40]

The carboxy-functionalized solid supports were obtained as outlined in Scheme 1. 10-Undecenoic acid (1) was converted into its tertiarybutyldimethylsilyl ester (2) by treatment with t-butyldimethylchlorosilane and 2 was hydro-

Scheme 1 : Preparation of the ω-carboxyalkylsilanized solid supports **5**

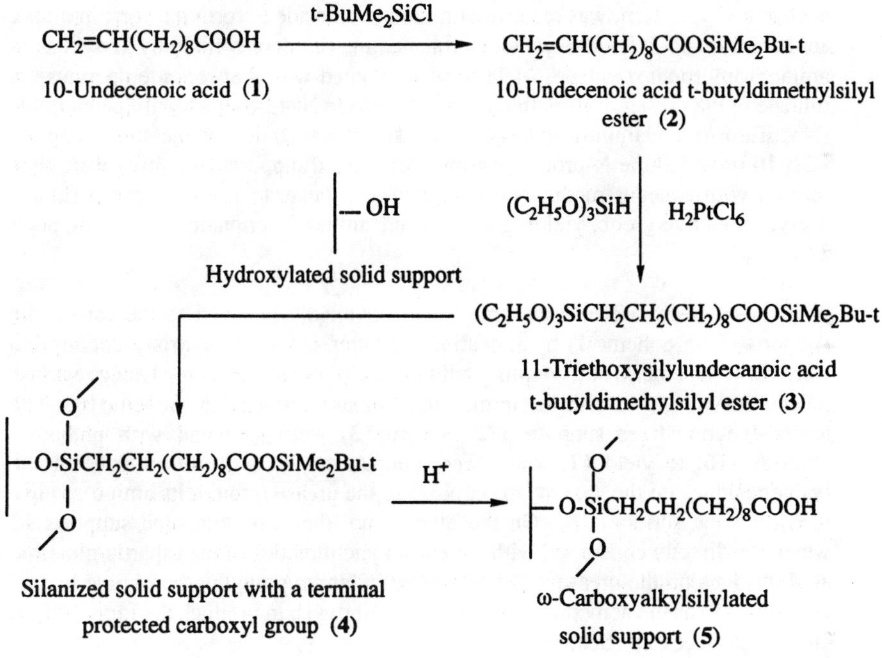

$$\underset{\text{10-Undecenoic acid (1)}}{CH_2=CH(CH_2)_8COOH} \xrightarrow{\text{t-BuMe}_2\text{SiCl}} \underset{\substack{\text{10-Undecenoic acid t-butyldimethylsilyl} \\ \text{ester (2)}}}{CH_2=CH(CH_2)_8COOSiMe_2Bu\text{-}t}$$

—OH $(C_2H_5O)_3SiH$ | H_2PtCl_6

Hydroxylated solid support

$$(C_2H_5O)_3SiCH_2CH_2(CH_2)_8COOSiMe_2Bu\text{-}t$$

11-Triethoxysilylundecanoic acid
t-butyldimethylsilyl ester (3)

$$\underset{\substack{\text{Silanized solid support with a terminal} \\ \text{protected carboxyl group (4)}}}{\overset{O}{\underset{O}{\big|}}-O\text{-}SiCH_2CH_2(CH_2)_8COOSiMe_2Bu\text{-}t} \xrightarrow{H^+} \underset{\substack{\omega\text{-Carboxyalkylsilylated} \\ \text{solid support (5)}}}{\overset{O}{\underset{O}{\big|}}-O\text{-}SiCH_2CH_2(CH_2)_8COOH}$$

silylated with triethoxysilane in the presence of chloroplatinic acid catalyst to afford 11-triethoxysilylundecanoic acid t-butyldimethylsilyl ester (**3**). The solid support was silanized with this reagent **3** and the silyl ester protective grouping on the silylated support **4** was removed by treatment with dilute hydrochloric acid, to furnish the desired carboxy-functionalized solid support **5**.

The amino-functionalized solid supports were obtained by a different procedure, as summarized in Scheme **2**. 11-Aminoundecanoic acid (**6**) was converted into its N-trifluoroacetyl derivative **7** by treatment with ethyltrifluoro-acetate and **7**, in turn, was reacted with thionyl chloride to form the corresponding acid chloride **8**. 3-Triethoxysilyl-1-propanamine (**9**, more commonly known as 3-aminopropyltriethoxysilane, APTES) was acylated with **8** at room temperature in toluene in the presence of triethylamine, to provide N^1-(3-triethoxysilylpropyl)-11-(N^2-trifluoroacetylamino) undecanamide (**10**). Silanization of the solid supports with **10** provided the N-protectedamino-derivatized supports **11**, which were then reacted with aqueous methanolic potassium carbonate to eliminate the trifluoro-acetyl protective group, yielding the desired amino-functionalized solid supports **12**.

The hydrolase enzyme urease was covalently attached to the carboxylic supports (**5**, in Scheme **1**) by activating the latter as their N-hydroxysuccinimido esters (**14**, Scheme **3**) and coupling with the ε-amino groups on the lysine residues on the enzyme, to furnish the immobilized urease surfaces **15** (Scheme **3**). The amino-functionalized supports (**12**, Scheme **3**) were activated with phthaloyl chloride (**16**) to yield **17**, which were coupled to the ε-amino moieties of the lysine residues on the enzyme to immobilize the urease through its amino groups, providing the surfaces **18**. On the other hand, the same aminated supports **12** were also directly condensed with the carboxylic moieties of the aspartic/glutamic acid residues on the urease using ethyldimethylaminopropyl carbodiimide hydro-chloride (**19**) as the activator of the enzyme carboxyls, to produce the immobilized enzyme surfaces **20** (Scheme **3**).

The immobilized urease surfaces in which the enzyme is bound to the functionalized solid supports through a phospholipid cross-linker were prepared as indicated in Scheme **4**. The amino groups of 1-palmitoyl-sn-glycerophospha-tidyl ethanolamine (**21**) were directly condensed with the N-hydroxysuccinimide-activated carboxylic supports (**14**) to give the covalently anchored lipid surfaces **22**. The free secondary hydroxyls at the sn-2 position of the lipid-bound surfaces **22** were acylated with the diacid chloride **23** and the resulting sn-2-ω-carboxyacyl-substituted lipid-attached surfaces **24** coupled to the ε-amino groups of the lysine residues on urease employing N-hydroxysuccinimide as the activator, furnishing the desired silanized support/lipid/urease surfaces **26** (Scheme **4**).

Alternately, the silanized support/lipid/urease matrices (**30**, Scheme **5**) were generated starting from the amino-functionalized solid supports **12**. These supports **12** were cross-linked to the lipid **21** using phthaloyl chloride (**16**) as the cross-linker to furnish the immobilized lipid surfaces **27** (Scheme **5**). Treatment of **27** with the acid chloride **8** and deprotection of the trifluoroacetyl groups on **28** with potassium carbonate provides the ω-amino-functionalized covalently-anchored

Scheme **2**: Formation of the ω-aminoacylaminopropylsilanized
solid supports (**12**)

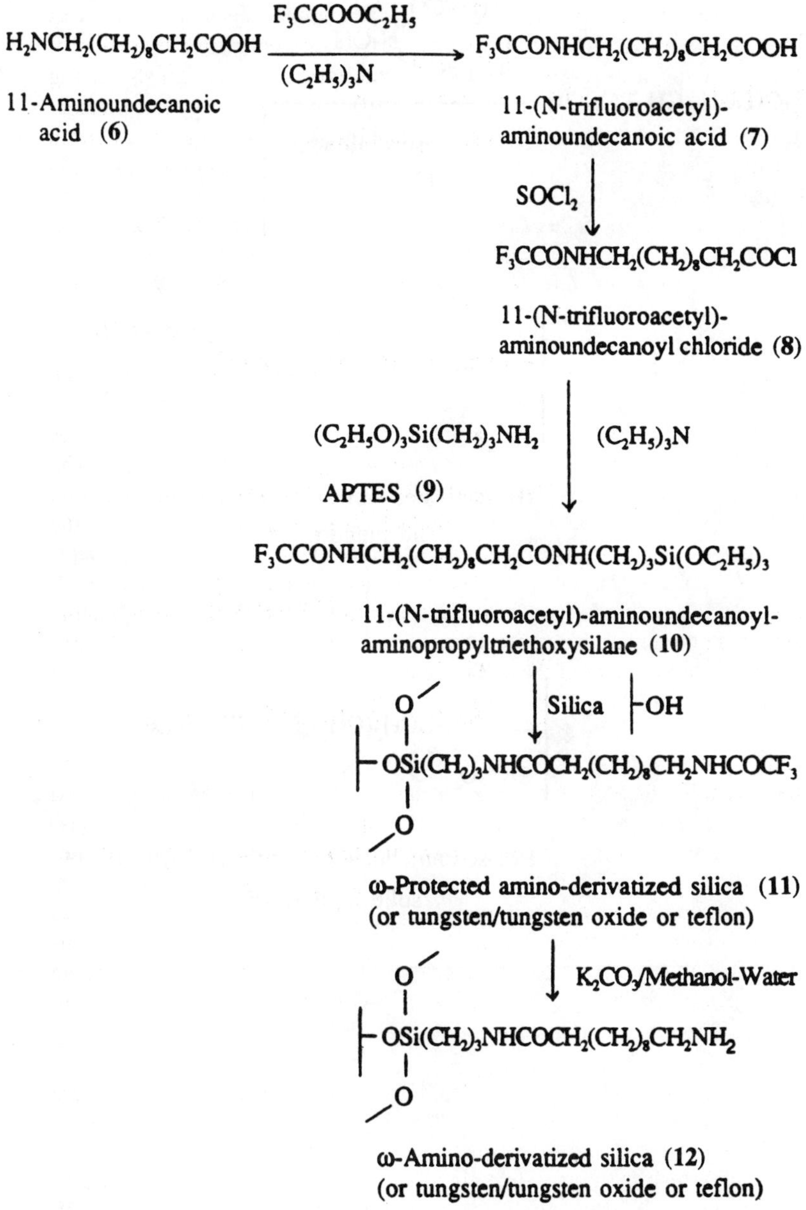

$$H_2NCH_2(CH_2)_8CH_2COOH \xrightarrow[\text{(C}_2\text{H}_5)_3\text{N}]{\text{F}_3\text{CCOOC}_2\text{H}_5} F_3CCONHCH_2(CH_2)_8CH_2COOH$$

11-Aminoundecanoic
acid (**6**)

11-(N-trifluoroacetyl)-
aminoundecanoic acid (**7**)

SOCl₂ ↓

$$F_3CCONHCH_2(CH_2)_8CH_2COCl$$

11-(N-trifluoroacetyl)-
aminoundecanoyl chloride (**8**)

$(C_2H_5O)_3Si(CH_2)_3NH_2$ ↓ $(C_2H_5)_3N$

APTES (**9**)

$$F_3CCONHCH_2(CH_2)_8CH_2CONH(CH_2)_3Si(OC_2H_5)_3$$

11-(N-trifluoroacetyl)-aminoundecanoyl-
aminopropyltriethoxysilane (**10**)

↓ Silica —OH

$$\left.\begin{array}{c}O \\ | \\ \vdash OSi(CH_2)_3NHCOCH_2(CH_2)_8CH_2NHCOCF_3 \\ | \\ O\end{array}\right.$$

ω-Protected amino-derivatized silica (**11**)
(or tungsten/tungsten oxide or teflon)

↓ K₂CO₃/Methanol-Water

$$\left.\begin{array}{c}O \\ | \\ \vdash OSi(CH_2)_3NHCOCH_2(CH_2)_8CH_2NH_2 \\ | \\ O\end{array}\right.$$

ω-Amino-derivatized silica (**12**)
(or tungsten/tungsten oxide or teflon)

Scheme 3 : Covalent binding of urease to the carboxy and amino-functionalized
silanized solid supports

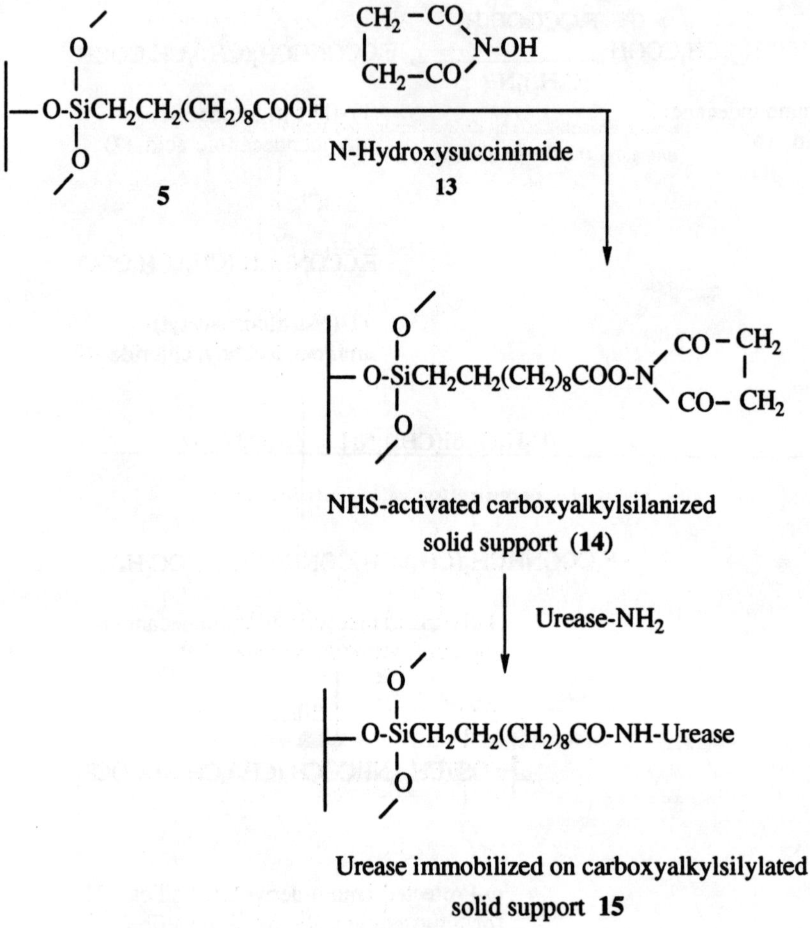

$$
\begin{array}{c}
\text{O} \\
| \\
-\text{O-SiCH}_2\text{CH}_2(\text{CH}_2)_8\text{COOH}
\end{array}
\quad
\xrightarrow[\text{N-Hydroxysuccinimide}]{
\begin{array}{c}
\text{CH}_2-\text{CO} \\
| \quad\quad \text{N-OH} \\
\text{CH}_2-\text{CO}
\end{array}
}
$$

5   13

$$
\begin{array}{c}
\text{O} \\
| \\
-\text{O-SiCH}_2\text{CH}_2(\text{CH}_2)_8\text{COO-N}
\begin{array}{c}
\text{CO}-\text{CH}_2 \\
| \\
\text{CO}-\text{CH}_2
\end{array} \\
| \\
\text{O}
\end{array}
$$

NHS-activated carboxyalkylsilanized
solid support **(14)**

Urease-NH$_2$

$$
\begin{array}{c}
\text{O} \\
| \\
-\text{O-SiCH}_2\text{CH}_2(\text{CH}_2)_8\text{CO-NH-Urease} \\
| \\
\text{O}
\end{array}
$$

Urease immobilized on carboxyalkylsilylated
solid support **15**

Scheme 3 (continued)

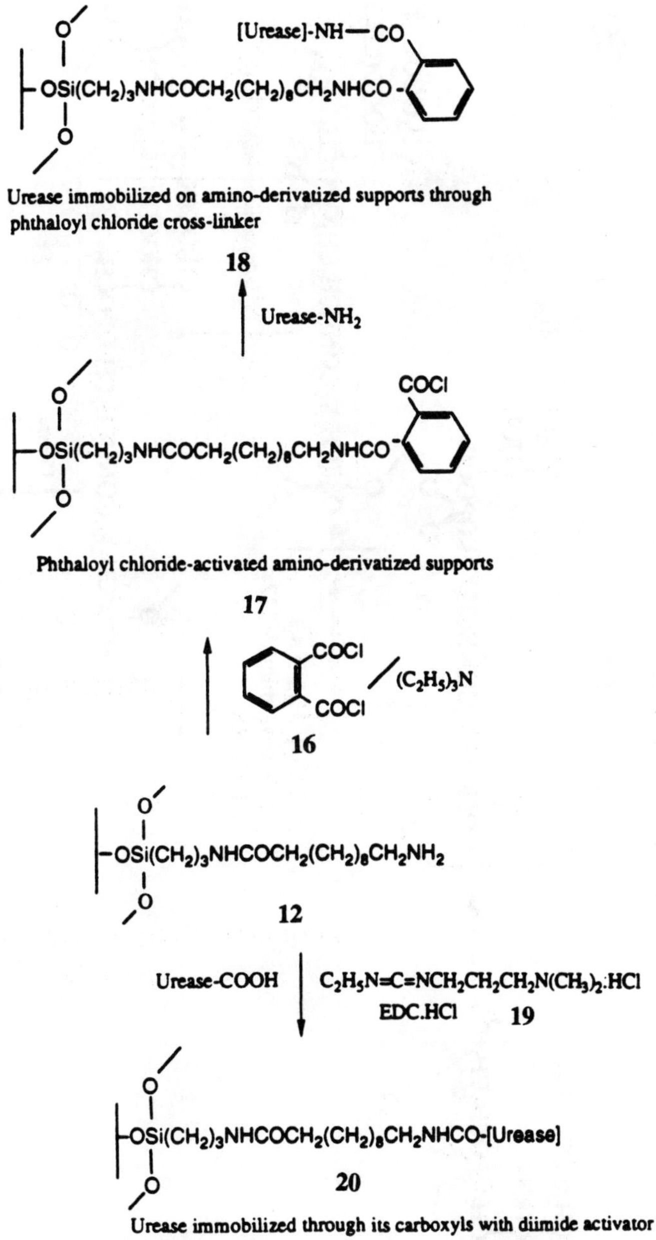

$[Urease]-NH—CO$

$-OSi(CH_2)_3NHCOCH_2(CH_2)_6CH_2NHCO-$

Urease immobilized on amino-derivatized supports through
phthaloyl chloride cross-linker

18

$\uparrow$ Urease-NH$_2$

$COCl$

$-OSi(CH_2)_3NHCOCH_2(CH_2)_6CH_2NHCO^-$

Phthaloyl chloride-activated amino-derivatized supports

17

$\begin{array}{c} COCl \\ COCl \end{array}$ $/(C_2H_5)_3N$

16

$-OSi(CH_2)_3NHCOCH_2(CH_2)_6CH_2NH_2$

12

Urease-COOH $\quad | \quad C_2H_5N=C=NCH_2CH_2CH_2N(CH_3)_2 \cdot HCl$

$EDC.HCl$ **19**

$-OSi(CH_2)_3NHCOCH_2(CH_2)_6CH_2NHCO-[Urease]$

20

Urease immobilized through its carboxyls with diimide activator

Scheme **4** : Covalent attachment of urease to carboxyalkylsilylated solid supports through phosphatidylethanolamine cross-linker

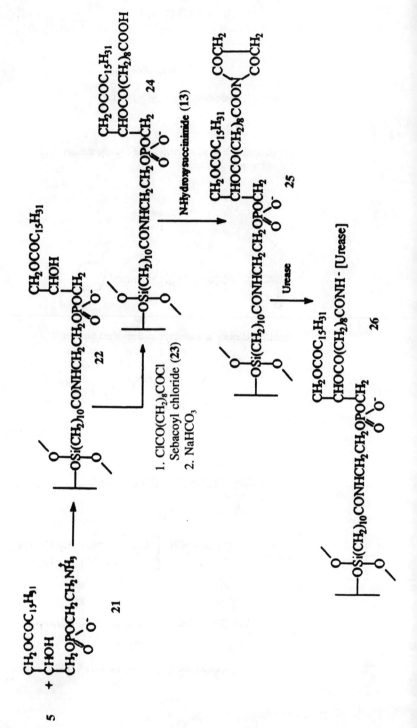

Scheme 5: Covalent immobilization of urease on aminoacyl-aminopropylsilylated solid supports

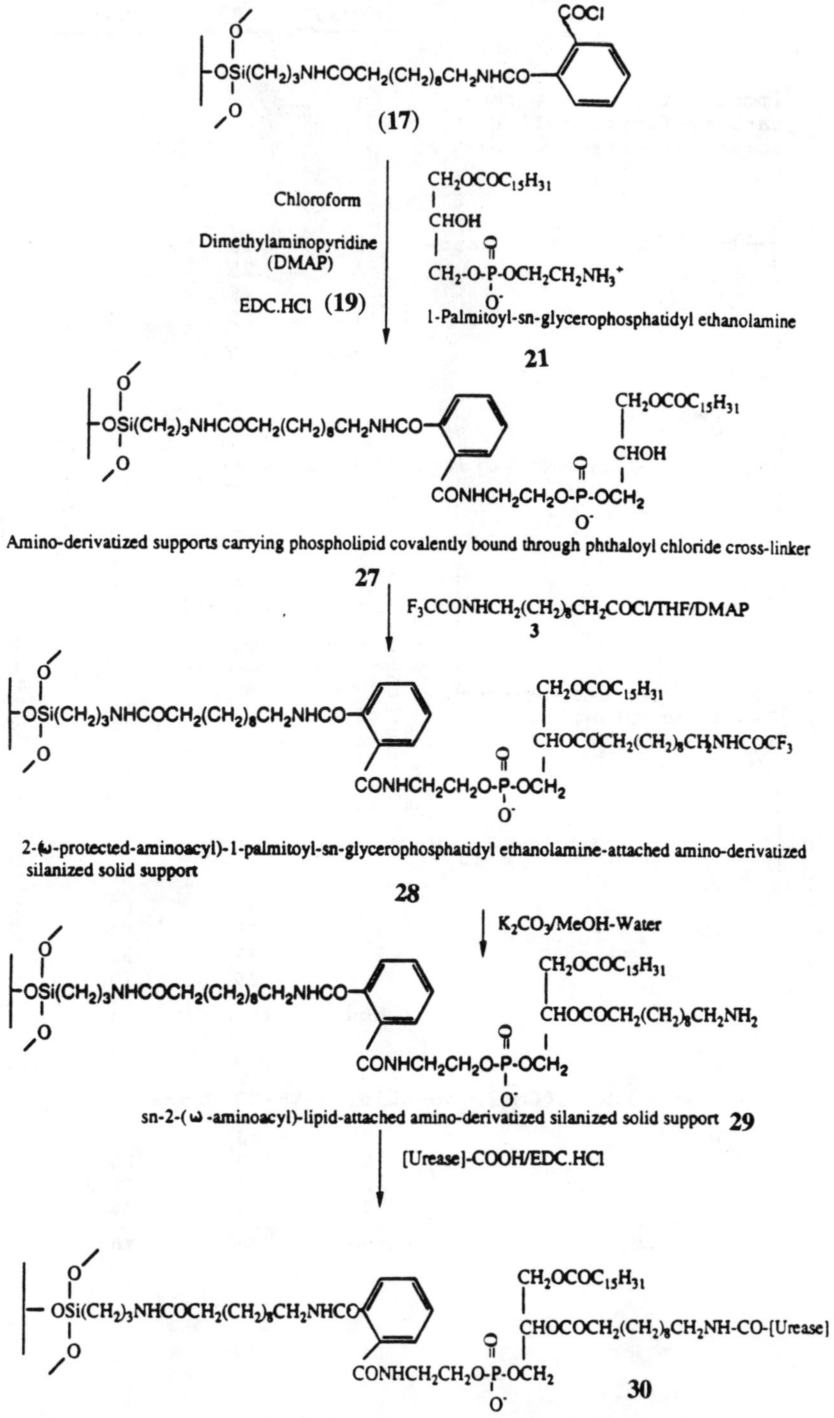

$$-OSi(CH_2)_3NHCOCH_2(CH_2)_8CH_2NHCO-$$

(17)

Chloroform

Dimethylaminopyridine (DMAP)

EDC.HCl (19)

$$CH_2OCOC_{15}H_{31}$$
$$CHOH$$
$$CH_2-O-P-OCH_2CH_2NH_3^+$$
$$O^-$$

1-Palmitoyl-sn-glycerophosphatidyl ethanolamine

21

$$-OSi(CH_2)_3NHCOCH_2(CH_2)_8CH_2NHCO-$$

$$CH_2OCOC_{15}H_{31}$$
$$CHOH$$
$$CONHCH_2CH_2O-P-OCH_2$$
$$O^-$$

Amino-derivatized supports carrying phospholipid covalently bound through phthaloyl chloride cross-linker

27

$$F_3CCONHCH_2(CH_2)_8CH_2COCl/THF/DMAP$$
3

$$-OSi(CH_2)_3NHCOCH_2(CH_2)_8CH_2NHCO-$$

$$CH_2OCOC_{15}H_{31}$$
$$CHOCOCH_2(CH_2)_8CH_2NHCOCF_3$$
$$CONHCH_2CH_2O-P-OCH_2$$
$$O^-$$

2-(ω-protected-aminoacyl)-1-palmitoyl-sn-glycerophosphatidyl ethanolamine-attached amino-derivatized silanized solid support

28

$$K_2CO_3/MeOH-Water$$

$$-OSi(CH_2)_3NHCOCH_2(CH_2)_8CH_2NHCO-$$

$$CH_2OCOC_{15}H_{31}$$
$$CHOCOCH_2(CH_2)_8CH_2NH_2$$
$$CONHCH_2CH_2O-P-OCH_2$$
$$O^-$$

sn-2-(ω-aminoacyl)-lipid-attached amino-derivatized silanized solid support 29

$$[Urease]-COOH/EDC.HCl$$

$$- OSi(CH_2)_3NHCOCH_2(CH_2)_8CH_2NHCO$$

$$CH_2OCOC_{15}H_{31}$$
$$CHOCOCH_2(CH_2)_8CH_2NH-CO-[Urease]$$
$$CONHCH_2CH_2O-P-OCH_2$$
$$O^-$$

30

Table 1:   Ultraviolet Spectrophotometric Assay of the
enzymatic activity of the immobilized urease surfaces

Surface	Storage time (days)	Temp. (°C)	% Initial activity retained

Immobilized urease from
carboxy-functionalised
solid supports:

$\vdash$O-Si(CH$_2$)$_{10}$CONH-Urease

	0	25	78
	1	25	12
	7	25	12
	42	25	12
	1 Hour	100	19

$\vdash$O-Si(CH$_2$)$_{10}$CO-NH-[Lipid]-CO-NH-Urease

	0	25	91
	1	25	88
	7	25	88
	42	25	88
	1 Hour	100	88

Immobilized urease from
amino-functionalised
solid supports:

$\vdash$O-Si(CH$_2$)$_3$NHCO(CH$_2$)$_{10}$NH-CO-Urease

	0	25	60
	10	25	51
	21	25	36
	42	25	24
	1 Hour	100	33

$\vdash$O-Si(CH$_2$)$_3$NHCO(CH$_2$)$_{10}$NH-[Lipid]-NH-CO-Urease

	0	25	74
	10	25	74
	21	25	74
	42	25	74
	1 Hour	100	74

lipid surfaces **29**. Condensation of the amino groups on **29** with urease using ethyldimethylaminopropylcarbodiimide hydrochloride activator yielded the desired matrices **30**.

All of the immobilized urease surfaces (with or without lipid cross-linkers) were characterized by ellipsometry and X-ray photoelectron spectroscopy, details of which can be found elsewhere.[43,44]. Immobilization yields were found to range from 45% to 70%, the procedures utilizing the carboxylic moieties of urease affording the highest values.

Enzymatic activity assays on all the immobilized urease matrices were carried out by ultraviolet spectrophotometric measurements of the absorption at 586 nm of the purple coloured solutions formed by the treatment of 1 mg of each of the matrices with the substrate solution consisting of urea and the pH-sensitive dye bromocresol purple. The enzyme reacts with the urea in the substrate solution liberating ammonia, with bicarbonate as the by-product. The results included in Table **1** indicate that the lipid-containing surfaces retained almost the same activity as was measured for these matrices immediately following immobilization, even after six weeks at room temperature. It should be recollected that these immobilized enzyme matrices were stored dry at room temperature in sample vials. On the other hand, the non-lipid immobilized enzyme surfaces lose 50-70% of their activity during the same storage period. Furthermore, when the lipid-bound enzyme matrices were heated either in the dry state in the oven at 100°C, or in boiling water for an hour, no significant loss of activity was noticeable. However, the non-lipid enzyme surfaces lost their activity completely under these conditions.

Our results with urease clearly demonstrate that thermal stabilities higher than ever recorded for any enzyme (including the thermophilic enzyme species) could be achieved by combining the rigidity generated through immobilization with the stabilizing effect of the natural surfactant phospholipid molecules. The overall configuration of the immobilized silanized solid support/lipid/enzyme matrix is structurally similar to that of the phospholipid reverse micelles/enzyme system[41], the difference being that the former is heterogeneous with part of the hydrophobic zone stemming from the silane and the rest from the lipid. The fact that rigidification of the enzyme by covalent immobilization by itself is not enough to impart a high degree of thermal stability is evident from the rates of acceleration of the catalysis reaction of urea hydrolysis by urease which is six-fold higher for the lipid-containing matrix in comparison with its non-lipid counterpart at 60°C, as determined by ultraviolet spectrophotometry at that temperature.

REFERENCES

1.   A.P.F.Turner (Ed.), 'Advances in Biosensors', Volume 1 (1991) and 2 (1992), Jai Press, London.
2.   I.Karube and H.Endo: Biosensors in Biodegradation of Wastes, in 'Biological Degradation of Wastes', A.M.Martin (Ed.), Elsevier, London, 1991, pp.103-132.

3. J.H.Fendler, 'Membrane-mimetic approach to Advanced Materials', Springer-Verlag, Heidelberg (in press).

4. 'Proceedings of the 7th Working Conference on Applied Surface Analysis', in Fresenius J.Anal.Chem. 1993, 346, 1-388.

5. A.E.Ivanov, V.V.Saburov and V.P.Zubov: Polymer-coated Adsorbents for the Separation of Biopolymers and Particles, in 'Advances in Polymer Science', Vol.104, pp.135-176, Springer/Verlag, Berlin, 1992.

6. A.van Blaaderen and A.Vrij, Langmuir, 1992, 8, 2921-31.

7. T.G.Bee, E.M.Cross, A.J.Dias, K.Lee, M.S.Shoichet and T.J.McCarthy, J.Adh.Sci.Technol., 1992, 6, 719-31.

8. J.Jagur-Grodzinski, Prog.Polym.Sci., 1992, 17, 361-415.

9. J.E.Sandoval and J.J.Pesek, Anal.Chem., 1991, 63, 2634-2641.

10. M.Lazar, R.Rado and J.Rychly, 'Advances in Polymer Science', Vol.95 (Polymer Physics), pp.149-197, Springer-Verlag, Berlin, 1990.

11. a) L.M.Siperko and R.R.Thomas, J.Adh.Sci.Technol., 1989, 3, 157-173.
    b) T.J.McCarthy, Chimia, 1990, 44, 316-18.

12. D.Chapman, Langmuir, 1993, 9, 39-45.

13. C.G.L. Khoo, J.B.Lando and H.Ishida, J.Polym.Sci. Part B: Polymer Physics, 1990, 28, 213-232.

14. M.Wolpers, M.Stratmann and H.Viefhaus, Fresenius J.Anal.Chem., 1991, 341, 337-8.

15. M.Mahon, K.W.Wulser and M.A.Langell, Langmuir, 991, 7, 486-92.

16. a) W.C.Feist, For.Prod.J., 1990, 40, 21-6.
    b) N.V.Velikanova, G.N.Myshelova, E.N.Pokrovskaya and V.I.Sidorov, Khim.Drev., 1989, (6), 100-3; Chem.Abstr., 1990, 112, 181684.

17. M.Takagi, T.Norimatsu, T.Yamanaka, S.Nakai and H.Ito, J.Vac.Sci. Technol. A, 1991, 9, 820-3.

18. M.Salkind, Angew.Chem.Int.Ed.Engl.Adv.Mater., 1989, 28, 655-662.

19. N.R.Lerner and T.Wydeven, J.Appl.Polym.Sci., 1989, 37, 3343-3355.

20. M.Alvarez-Icaza and U.Bilitewski, Anal.Chem., 1993, 65, 525A-533A.

21. J.Janata, Anal.Chem., 1992, 64, 196R-219R.

22. a) U.Pyell and G.Stork, Fresenius J.Anal.Chem., 1992, 342, 281-86, 376-380.
    b) C.Ferrari, G.Predieri and A.Tiripicchio, Chem.Mater., 1992, 4, 243-45.

23. P.Hernan, C.del Pino and E.Ruiz-Hitzky, Chem.Mater., 1992, 4, 49-55.

24. a) 'Biodegradable Polymers', in 'Encyclopedia of Polymer Science and Engineering, Vol.2, pp.220-243, Wiley, New York, 1985.
    b) 'Drug and Enzyme Targeting', in 'Methods in Enzymology', Vol.149, Part B, R.Green and K.J.Widder (Eds.), Academic, San Diego, 1987.

25. L.A.Harrison: Immobilized Enzymes and their applications, in 'Biotechnology of Waste Treatment and Exploitation', J.M.Sidwick and R.S.Holdom (Eds.), Ellis Harwood Ltd., Chichester, 1987, Chapter 4, pp.81-120.

26. W.Boland, C.Frobl and M.Lorenz, Synthesis, 1991, 1049-1072.

27. H.J.M.Op Den Camp and H.J.Gijzen, in 'Biological Degradation of Wastes', A.M.Martin (Ed.), Elsevier, London, 1991, pp.281-306.

28. P.Vadgama and P.W.Crump, Analyst, 1992, 117, 1657-1670.

29. B.Fleet and H.Gunasingham, Talanta, 1992, 39, 1449-1457.

30. G.Rialdi, E.Battistel, L.Benatti and P.Sabbioneta, J.Thermal Analysis, 1992, 38, 159-167.
31. L.Gianfreda and M.R.Scarfi, Mol.Cell.Biol., 1991, 100, 97-128.
32. M.N.Gupta, Biotechnol.Appl.Biochem., 1991, 14, 1-11.
33. R.F.Taylor, in 'Protein Immobilization: Fundamentals and Applications', R.F.Taylor (Ed.), Marcel Dekker, New York, 1991, p.2.
34. J.M.S.Cabral and J.F.Kennedy, in 'Protein Immobilization: Fundamentals and Applications', R.F.Taylor (Ed.), Marcel Dekker, New York, 1991, pp.73-179.
35. R.J.Markovich, X.Qiu, D.E.Nichols, C.Pidgeon, B.Invergo and F.M.Alvarez, Anal.Chem., 1991, 63, 1851-1860.
36. L.Haussling, B.Michel, H.Ringsdorf and H.Rohrer, Angew.Chem.Int.Ed. Engl., 1991, 30, 569-572 (and references therein).
37. S.J.Vigmond, K.M.R.Kallury and M.Thompson, Anal.Chem., 1992, 64, 2763-2769 (and references cited therein).
38. D.J.Hook, T.G.Varago, J.A.Gardella,Jr., K.S.Litwiler and F.V.Bright, Langmuir, 1991, 7, 142-151.
39. W.J.van Ooij and R.S.Michael, ACS Symposium Series 440: Metallization of Polymers, E.Sacher, J.Pireaux and S.P.Kowaalczyk (Eds.), American Chemical Society, 1990, Chapter 4, pp.60-87.
40. N.B.McKeown, P.G.Kalman, R.Sodhi, A.D.Romaschin and M.Thompson, Langmuir, 1991, 7, 2146-2152.
41. A.Darszon and L.Shoshani, 'Enzymes in Reverse Micelles containing Phospholipids', in 'Biomolecules in Organic Solvents', A.Gomez-Puyou (Ed.), CRC Press, Boca Raton, 1992, pp.35-66.
42. M.Przybyt and H.Sugier, Anal.Chim.Acta, 1990, 239, 269-276.
43. K.M.R.Kallury, W.E.Lee and M.Thompson, Anal.Chem., 1992, 64, 1062-68.
44. K.M.R.Kallury, W.E.Lee and M.Thompson, Anal.chem., (in press).

Ellipsometry, X-Ray Photoelectron Spectroscopy, and Surface Plasmon Resonance as Techniques for the Study of Chemically Modified Surfaces

John D. Brennan, R. F. De Bono, Krishna M. R. Kallury, and Ulrich J. Krull*

CHEMICAL SENSORS GROUP, ERINDALE CAMPUS, UNIVERSITY OF TORONTO, 3359 MISSISSAUGA RD. N., MISSISSAUGA, ONTARIO, L5L 1C6, CANADA

1 INTRODUCTION

Chemical modification of surfaces by immobilization of ultra-thin organic films of surfactants has been of interest in areas such as surface passivation and bio-compatibility. It has been reported that a layer of phospholipids or long-chain fatty acids or amines may reduce the rate of denaturation of proteins that are immobilized on such a surface.[1] Membranes composed of covalently immobilized amphiphiles and proteins have been prepared, but are usually poorly characterized with respect to structure and function in real time and *in situ*.[2]

The need for characterization of physical structure and chemical speciation of chemically modified surfaces has prompted the development of a number of analytical methods that are suitable for investigations of monolayers of organic films. A standard technique for characterization of organic monolayer films is x-ray photoelectron spectroscopy (XPS). XPS can provide information on the chemical speciation, packing density and thickness of organic overlayers on solid substrates such as metals, quartz or silicon.[2] This technique requires that the sample be present in ultra-high vacuum (UHV), and also requires that the sample be irradiated with x-ray radiation over an area of several square millimeters for extended periods of time. While this technique is generally described as "non-destructive", application of XPS to the study of immobilized proteins often results in substantial decreases in the biochemical activity of these species. The use of UHV, in combination with sample damage which may result from irradiation with x-rays, can alter the chemical and physical structure of the film and therefore may result in inaccurate information. The need to place the sample into UHV before collecting data also removes the possibility of using XPS to measure dynamic events such as adsorption of species onto a surface. XPS can only provide a static "snapshot" of the structure of a monolayer, and can not be used to follow events in real-time.

Alternative techniques for the characterization of organic films include ellipsometry and surface infrared and Raman spectroscopy. These techniques do not require UHV conditions and therefore do not damage the sample. Ellipsometry involves the measurement of changes in the polarization of light upon reflection from a surface which is coated with an overlayer, and provides information about the thickness and refractive index of monolayer films. The need for reflection necessitates that the monolayers be present on substrates such as metals or silicon. Ellipsometry is an averaging technique that samples areas on the order of 1 mm^2, and can measure changes in average thickness of 0.1 nm.[3] Ellipsometry is not commonly used to measure dynamic events which are occurring at surfaces.

A relatively new technique which can operate *in situ* is surface plasmon resonance (SPR) spectroscopy. In the Kretschmann configuration the base of a prism is coated with a thin layer of metal, and parallel polarized light is passed into the prism at an angle of incidence greater than the critical angle (that is defined by the prism and the outer medium) to reflect from the metal/prism boundary. A minimum in reflected intensity from the boundary is observed at a specific angle due to the excitation of plasmons in the conduction band of the metal.[4] The resonant excitation of the collective motion of electrons in the conduction band produces an associated evanescent wave which decays into the outer medium. The resonance condition (i.e. angle for absorption of light) is influenced by the thickness and refractive index of any organic film in place between the metal and the outer medium. The ability to operate *in situ* allows for the measurement of adsorption of organic films onto metal surfaces in real time.[5] The non-perturbing nature of SPR spectroscopy also suggests that sample damage should be minimized, and therefore species such as proteins should remain active during and after SPR measurements.

The techniques described above provide information about structure over relatively large areas (square millimeters). Information about the microscopic organization and structure of monolayers of amphiphiles prepared by Langmuir-Blodgett techniques, and visualization of such structures, may be achieved by fluorescence microscopy.[6,7] This method makes use of a fluorescent dye which can distribute among ordered and disordered phases within a monolayer so that solid areas show little fluorescence intensity and fluid areas contain substantially greater quantities of the dye and are therefore relatively bright. Fluorescence microscopy provides information on the basis of the distribution of a dye within a mixed phase system, but partition coefficients are poorly defined and the use of dye has the potential to perturb the structure of films which are investigated. In addition, the lack of fluorescence emission close to conductive surfaces indicates that this technique is applicable only to monolayers which are present on substrates such as quartz or glass.

SPR has recently been extended to provide a newer method known as surface plasmon microscopy (SPM), which can provide information about gradients of refractive index or thickness with a spatial resolution of about 5 μm (coherence length of plasmon) in the form of an image of variable intensity.[8-10] The surface plasmon resonance condition is determined by the average of the thickness and refractive index of the organic film as sampled over the coherence length near a metal surface. Differences in the profile of the film over distances greater than the coherence length will be detected as spatially resolved differences in reflected intensity at a fixed angle of incidence. This forms the basis of surface plasmon microscopy which uses these differences in intensity to provide optical contrast to image regions of different optical mass (refractive index, thickness).

This work reports the immobilization and study of long-chain carboxylic acids by silane coupling to quartz or silicon substrates, self-assembly of long-chain thiols onto metallic gold substrates, and subsequent attachment of active proteins to both of these coatings. Ellipsometry and XPS were used to determine the structure of the silane-based films on quartz and silicon, respectively, and the extent of coverage of substrates with amphiphiles and proteins. Surface plasmon resonance was used to investigate the *in situ* dynamic properties of the thiol-based films on gold, as well as the kinetics of protein adsorption to these films. The information available from XPS, ellipsometry, SPR and SPM techniques is compared, and the applicability of these techniques to the study of protein monolayers is discussed.

2 EXPERIMENTAL

Chemicals

Urease (EC 3.5.1.5) type IV from jack bean (activity of 105,000 Unit.g^{-1}), concanavalin A type IV and urea were purchased from Sigma Chemical Company (St. Louis, MO). 16-hydroxy-hexadecanoic acid, carbonyldiimidazole (CDI), dimethyldichlorosilane, hexadecylmercaptan and chloroplatinic acid were purchased from Aldrich Chemical Company (Milwaukee, WI). Dextran T500 was from Pharmacia (Uppsala, Sweden). All other chemicals were of reagent grade.

The quartz wafers and silicon wafers were purchased from Heraeus-Amersil Inc. (Sayerville, NJ). Gold (99.99 % purity) was from Deak International (Mississauga, ON) and chromium was from R.D. Mathis (Long Beach, CA).

Equipment

X-ray photoelectron spectroscopy (XPS) of covalently stabilized membranes was done using a Leybold MAX-200 spectrometer (Leybold-Heraeus, Cologne, FRG) with excitation by unmonochromatized Mg K_α radiation and a spot size of 2 x 4 mm^2. An excitation voltage of 1253.6 eV, a detection voltage of 2.65 eV and an emission current of 20 µA were utilized. Pass energies of 192 eV and 48 eV were used for broad and narrow region scans respectively. A take-off angle of 90° was used for both broad region and narrow region scans, unless otherwise stated. The intensities reported were corrected for Scofield factors and instrument transmission factors using software routines which were provided by Leybold. The shapes of the peaks indicated that no compensation for differential surface charging was necessary. Satellite peaks which resulted from the unmonochromatic Mg K_α radiation were subtracted from all spectra. The binding energy scale of the spectrometer was calibrated to the Ag($3d_{5/2}$) and Cu($2p_{3/2}$) peaks at 368.3 eV and 932.7 eV respectively, and the binding energy scale was shifted to place the C(1s) peak at 285.0 eV.

Ellipsometry of monolayers which were immobilized onto silicon wafers was done with a Rudolph Research AutoEL-II null reflection ellipsometer (Rudolph Research Corp., Flanders, NJ) utilizing a wavelength of 632.8 nm and an incident angle of 70°. Measured values of Δ and Ψ were converted to thickness and refractive index information using software originally developed by McCracken.[11]

Infrared spectra of membrane precursors were collected using a modified Bomem Fourier Transform Infrared (FTIR) spectrometer operated in the transmittance mode.[12]

The apparatus used to prepare metal-coated glass slides was a Key high vacuum metal vapour deposition unit (Key High Vacuum Products, Nesconset, NY).

The experimental apparatus used for surface plasmon resonance studies is shown in Fig.1. Angular control was achieved by use of a goniometer from a Rudolph Research ellipsometer. The optical train consisted of a 2 mW linearly polarized helium-neon laser (Spectra-Physics, Carlsbad, CA) which emitted at 632.8 nm with a beam diameter of 2 mm, a divergence of 0.01 mradians and a stability better than 0.8 %; a dichroic polarizer (Melles Griot, Nepean, ON) and two Glann Thompson polarizers. The laser beam was reflected from the base of the prism to a 5x microscope objective lens which was positioned such that the focal point was at the base of the prism. The reflected light was collected by the lens and transmitted to a Hitachi KP-111 CCD solid state camera (automatic gain disabled).

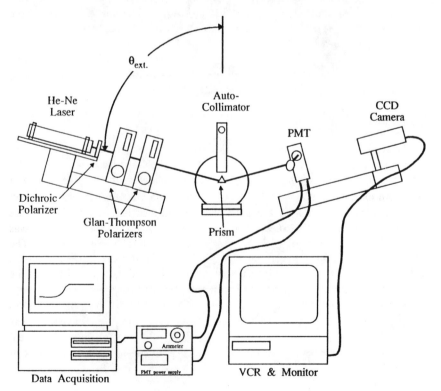

Figure 1. Surface plasmon resonance spectrometer and microscope.

Procedures

Cleaning of Substrates. All substrates (quartz and silicon wafers and glass microscope slides) were washed in hot detergent for 30 minutes, and were then sonicated for 30 minutes. The wafers were washed in distilled water, soaked in chromic acid for 10-15 minutes and rinsed thoroughly with distilled water. The wafers were then dried for at least 30 minutes in an oven at 110 °C.

Preparation of Silane Membranes with Immobilized Urease. Urease was adsorbed onto silicon wafers by incubating the wafers in a solution consisting of 1 mg.ml^{-1} of urease in phosphate buffer (pH 7.4) for a period of 12 hours. The extent of coverage was determined by ellipsometry assuming a refractive index of 1.50 for the coating.[13] A minimum of 5 spots were measured to provide an average thickness value for each sample.

Covalent immobilization of urease onto C-16 spacer chains began with the preparation of covalently immobilized membranes with acyl chains consisting of 16 carbons and containing carboxylic acid headgroups. The multi-step synthesis is described below and shown in Figure 2.

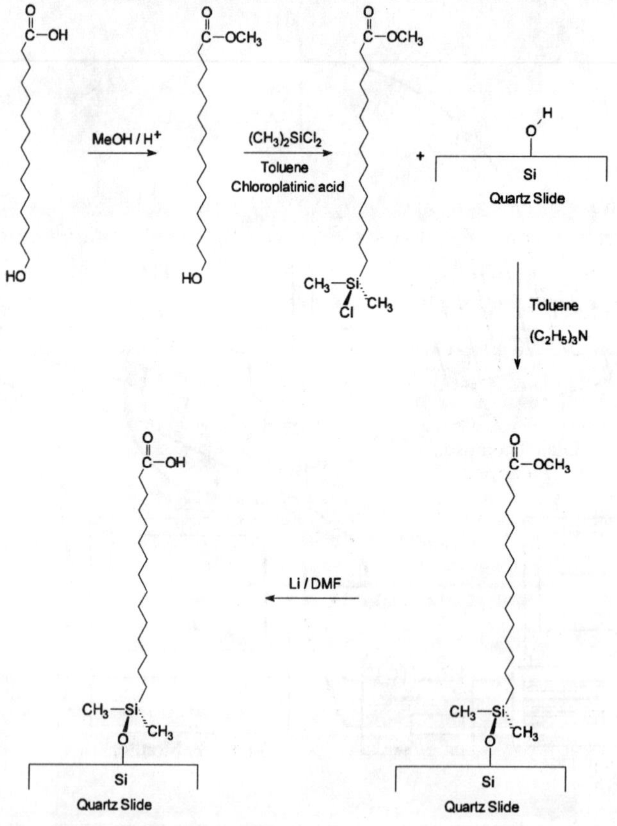

Figure 2. Outline of the synthetic procedure used to immobilized hexadecanoic acid membranes onto quartz surfaces.

16-OH hexadecanoic acid (1 millimole) was refluxed in a solution of 10 ml of methanol and 0.5 ml of concentrated hydrochloric acid for five hours under anhydrous conditions. The reaction mixture was then poured into a 100 ml solution of 2% sodium bicarbonate at pH 8. An extraction with ether was carried out and the ether layer was washed thoroughly with water, dried over anhydrous magnesium sulfate and concentrated in a rotary evaporator. The resulting methyl ester of 16-OH hexadecanoic acid was obtained in almost 100% yield. The purity of the methyl ester was checked using FTIR and ^{1}H-NMR.

The methyl ester (1 millimole) was dissolved in 25 ml of a mixture of dry toluene containing 1.1 millimoles of dimethyldichlorosilane and 10 mg of chloroplatinic acid catalyst. This mixture was refluxed for 16 hours. The solvent and unreacted silane were removed under vacuum. The mixture was then distilled under vacuum and the fraction boiling at 140-145°C at a pressure of 1 mm Hg was collected.

Clean quartz wafers were suspended in 10 ml of a 1% solution of 16-(dimethylchlorosilyloxy) hexadecanoic acid methyl ester in toluene. A few drops of triethylamine catalyst was added and the mixture was allowed to stand under nitrogen at room temperature overnight. The wafers were removed and washed thoroughly with dichloromethane and dried in a vacuum dessicator.

The quartz wafers were suspended in 10 ml of dimethylformamide and 500 mg of lithium iodide was added. This mixture was refluxed under nitrogen for 16 hours after which the wafers were washed with water, methanol, chloroform, and then acetone. The wafers were then dried in a vacuum dessicator.

Urease was immobilized onto the carboxylic acid membranes by activating the membrane with carbonyl diimidazole (CDI) prior to immobilization of urease. This reaction scheme is shown in Figure 3. The substrate was suspended in THF (5 ml) and 20 mg of CDI was added. The substrate was allowed to stand at room temperature for 2 hours and was then washed with THF. The substrate was suspended in water (1 ml) and urease (1 mg) was added. The mixture was allowed to stand for 12 hours at 5 °C. The substrate was washed with water and stored at minus 20°C until used.

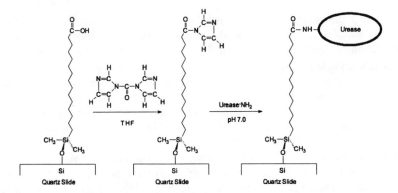

Figure 3. Outline of the synthetic procedure used to immobilize urease onto hexadecanoic acid membranes.

Preparation of Thiol Monolayers with Concanavalin A . The glass microscope slides were first coated with 3 nm of chromium (to improve adhesion of gold) and then 40 nm of gold by vacuum deposition. The gold coated glass slide was used as one wall of a flow-through teflon cell, and was optically connected to the 60° BK-7 prism by matching immersion oil (Type B, R.P. Cargille Laboratories). Continuous solution flow from a peristaltic pump was used to deliver solutions of hexadecylmercaptan in ethanol (10 µM concentration) or aqueous solutions of concanavalin A (to 85 nM concentration), dextran (to 1.46 µM), or Ca^{2+} and Mn^{2+} ions (to 100 µM) to the solution cell.

3 RESULTS AND DISCUSSION

Ellipsometry Studies of Alkylsilane Membranes with Immobilized Urease

All samples used for ellipsometric measurements consisted of covalently immobilized membranes (with or without proteins) on silicon wafers. It was not possible to use quartz surfaces since these are only partially reflective and thus do not permit accurate determination of optical constants for the substrate. In addition, the refractive index of quartz (1.47) is similar to that of many biological coatings, and thus substantial reflection of radiation at the film-substrate interface may not occur. The refractive index of silicon is 3.84, and therefore this problem does not exist for silicon.

The surface of silicon is similar to quartz in that there is a natural layer of oxidized silicon. The procedure which was used to clean the silicon slides involved incubation in chromic acid, which results in the presence of hydroxyl groups at the surface of the silicon. These groups provide anchoring points which allow for attachment of chlorosilane species. The density and distribution of hydroxyl groups is likely very different for quartz and silicon, and it is also possible that the surface roughness of quartz (as compared to silicon) will result in a greater overall number of hydroxyl groups. This may affect the final density of covalently immobilized membranes such that the quartz and silicon surfaces have slightly different amounts of amphiphile present at the surface.

Table 1 shows the ellipsometric data for the covalently immobilized membrane samples which were used in this work, both with and without immobilized urease. Calculation of thickness values used the following values for the various optical constants of the ambient-film-substrate system: wavelength of incident radiation (λ_e) = 632.8 nm, angle of incidence (Φ) = 70°, imaginary refractive index of ambient (k_0) = 0, real refractive index of air (n_0) = 1.0003, imaginary refractive index of coating (k_2) = 0 (assumes film is non-absorbing), real refractive index of coating (n_2) = 1.50, imaginary refractive index of substrate (k_s) = 0.2182 and real refractive index of substrate (n_s) = 3.8396 for the silicon substrate. The values of n_s and k_s were determined for each silicon wafer before immobilization of amphiphilic membranes. In all cases, the thickness values which were determined by ellipsometry were the average of thickness values which were measured randomly at 5 spots per sample for 3 samples.

Table 1. Ellipsometric data for samples studied.

Sample (See Figs. 2 and 3)	Thickness (nm)	Change in Thickness (nm) (Compared to membrane alone)
Urease adsorbed onto silicon	12.3 ± 1.6	-
C-16 carboxylic acid membrane	2.8 ± 0.1	-
C-16 carboxylic acid membrane + Urease immobilized by CDI	8.5 ± 0.1	5.7 ± 0.2

Ellipsometry of carboxylic acid membranes which were bound to silicon wafers showed the C-16 carboxylic acid membrane to be 2.8 ± 0.1 nm thick, which is in good agreement with the theoretical value of 2.8 nm calculated for a monolayer with an all-*trans* 16 carbon chain. The assumption of all-*trans* acyl chains is supported by the results of Sagiv *et al.*[14] who have done ATR-FTIR measurements of covalently immobilized membranes using polarized radiation. These measurements showed that in close packed monolayers the adsorbed moieties are oriented with their acyl chains perpendicular to the plane of the support.

Addition of urease to the carboxylic acid membrane by the CDI method caused the thickness of the membrane to increase by ca. 5.7 nm. The size of the urease can be computed from the mass and density of the enzyme, which are 476000 daltons and 1.40 g.cm^{-3}, respectively.[15] This corresponds to a radius of 5.2 nm for the enzyme, which suggests that the dimensions of the enzyme are roughly 10.0 nm x 10.0 nm x 11.0 nm. Ellipsometry of a monolayer of urease which was adsorbed to a silicon wafer from a solution of 1 mg.ml^{-1} of urease showed the thickness to be 12.3 ± 1.6 nm, in agreement with the calculated thickness of a monolayer of urease. The small increase in the thickness of urease immobilized to the carboxylic acid membrane (5.7 nm) when compared to the dimensions of urease (or to the thickness of a monolayer of adsorbed urease) indicates that there is incomplete coverage of the surface with urease. This thickness corresponds to a surface coverage of urease of approximately 50%.

It should be noted that thickness values which were determined by ellipsometry were *relative* values which were obtained by comparison of the thickness of bare silicon substrates and substrates which were covered with a covalently immobilized membrane or protein. It is extremely difficult to calculate absolute thickness values using ellipsometry because the real and imaginary refractive indices and the thickness of the film can not be accurately established. Furthermore, the covalently immobilized membrane system is not ideal for ellipsometry in that it violates several of the assumptions of the Drude equations.[3] For example, the film is oriented with acyl chains perpendicular to the air-film boundary and is therefore uniaxially anisotropic. In addition it is possible that the surfaces are not smooth and that the boundaries between ambient, film and substrate are not plane parallel, especially in the case where proteins are bound to the membrane. Finally, it is possible that the films are discontinuous and are present in the form of patchy overlayers.

Given the possible violations of the Drude equations, the thickness values obtained from ellipsometry are semi-quantitative at best. However, the relative ease and speed with which ellipsometric measurements can be done, in combination with the fact that the sample can be present in air, indicates that this technique is very useful for rapid measurements of film thickness. It can be concluded that the alkylsilane membranes investigated in this work were on average one monolayer in thickness, with a possible variation of $\pm$ 0.25 monolayers, while the protein layers were ca. 0.5 monolayers in thickness, based on the ellipsometric data.

X-ray Photoelectron Spectroscopy

A further method which was used for characterization of immobilized alkylsilane membranes and proteins was x-ray photoelectron spectroscopy. In this work, XPS was used to calculate the thickness (or surface coverage) of covalently immobilized organic overlayers which were present on quartz or silicon substrates. XPS spectra were collected at a take-off angle of 90° for the C-16 carboxylic acid membrane with and without immobilized urease, and for a monolayer of urease which had been adsorbed to a silicon wafer. The relative intensities of the C(1s), N(1s), O(1s) and Si(2p) peaks for each sample are given in Table 2. A broad scan XPS spectrum of the C-16 carboxylic acid membrane is shown in Figure 4a. A high resolution spectrum of the C(1s) region is shown in Figure 4b. A broad scan XPS spectrum of the C-16 carboxylic acid membrane with immobilized urease is shown in Figure 5a. A high resolution spectrum of the C(1s) region is shown in Figure 5b.

Table 2. X-ray photoelectron spectroscopic data for samples studied in this work.

Surface (See Figs. 2 and 3)	Elemental Composition				High Resolution Data					
	%C	%N	%Si	%O	C(1s) B_E	Area	N(1s) B_E	Area	Si(2p) B_E	Area
Urease on Si (Adsorbed)	64.8	14.3	0.4		285.0	47.0	400.3	100	99.0	
		20.5				100				
					286.4	29.8				
					288.2	23.2				
C-16 COOH on quartz	57.6	0.0	14.1		285.0	84.9			103.8	
		28.3				100				
					286.7	10.4				
C-16 COOH + Urease	70.4	9.1	2.1		288.7	4.7				
		18.4			285.0	65.5	400.4	85	103.5	
						100				
					286.4	22.4	402.1	15		
					288.5	11.6				

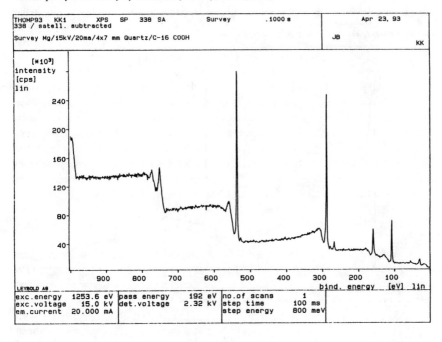

(A)

(B)

Figure 4. XPS spectra of hexadecanoic acid covalently immobilized on quartz. (A) broad scan spectrum, (B) narrow region spectrum of C(1s) binding energy region.

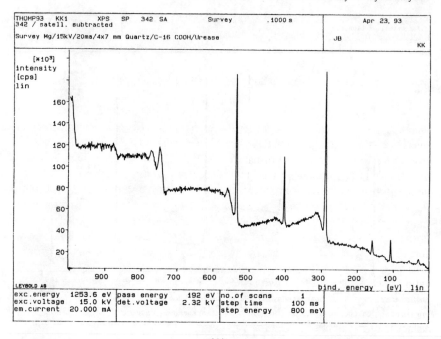

(A)

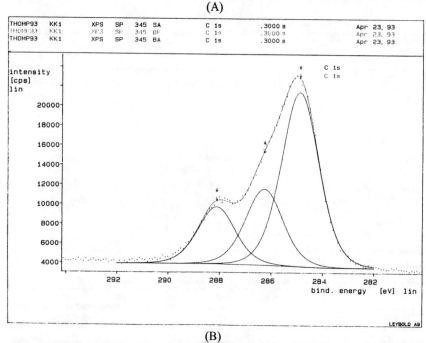

(B)

Figure 5. XPS spectra of urease covalently immobilized onto hexadecanoic acid membranes on quartz. (A) broad scan spectrum, (B) narrow region spectrum of C(1s) binding energy region.

In cases where XPS measurements are made at a single angle the intensity arising from a given elemental peak of the substrate (I_S) and the monolayer (I_M) may be used to calculate the thickness of the film using the following equation:[2]

$$\frac{I_M}{I_S} = \frac{\lambda_{MM}\{1 - \exp[-t_{MM}/(\lambda_{MM} \sin \theta)]\}}{\lambda_{SS}[\exp(-t_{SM}/\lambda_{SM} \sin \theta)}$$ (1)

where t_{MM} is the thickness of the overlayer through which an electron from the overlayer must travel, t_{SM} is the thickness of the overlayer through which an electron from the substrate must travel (usually $t_{MM} = t_{SM}$), λ_{MM} is the attenuation length for an electron from the monolayer travelling through the monolayer, λ_{SS} is the attenuation length for an electron from the substrate travelling through the substrate, λ_{SM} is the attenuation length for an electron from the substrate travelling through the monolayer and θ is the angle of emission of the electron relative to the surface normal (take-off angle). Equation (1) assumes that the spectra have been corrected for differences in ionization cross-section and instrument transmission efficiency which exist for electrons from the monolayer and substrate. The calculation of a ratio of intensity values from two peaks within a single spectrum removes problems which may be associated with changes of incident x-ray intensity, and is thus a better method for calculating the thickness of overlayers than comparison of intensity values from two different spectra.

In cases where data is collected at a single angle, it is best to collect at an angle of 90° so that sufficient signal-to-noise ratios will be present for both the substrate and overlayer, providing more accurate calculations of overlayer thickness.

In the case of the C-16 carboxylic acid membrane (with or without urease) the substrate used was quartz, which is composed of silicon and oxygen. The overlayer is composed of silicon, oxygen and carbon (and nitrogen if urease is present). The silicon of the substrate does not have the same binding energy as the silicon of the monolayer, since the silicon of the monolayer is a dimethylsilane, which has a binding energy of 102.5 eV, while the silane of the quartz is SiO_2, which has a binding energy of 103.9 eV. This indicates that unique signals exist which can be attributed to the monolayer (the C(1s) peak) and the substrate (the Si(2p) peak at 103.9 eV). The peaks may be used in the calculation of overlayer thickness values.

Equation (1) was used to calculate the thickness of the C-16 carboxylic acid monolayer using the following values: $\gamma_{MM} = 3.9$ nm, $\gamma_{SS} = 3.6$ nm, $\gamma_{SM} = 3.6$ nm,[2] and $\theta = 90°$. The average thickness was 5.3 ± 0.5 nm. It should be noted that the use of the C(1s) peak will introduce some inaccuracy into the thickness value owing to the presence of organic contamination. Airborne organic matter is known to adsorb to both hydrophilic and hydrophobic surfaces. In fact, it has been shown that an ultra-clean surface prepared in ultrahigh vacuum will become coated with organic contaminants in a matter of seconds upon exposure to air, and that such contaminants are not removed when the sample is placed back in an ultrahigh vacuum environment.[16] An XPS spectrum of bare quartz was collected to determine the thickness of the contaminant layer. The thickness was calculated to be on the order of 1.0 ± 0.2 nm, indicating that the thickness values calculated for the self-assembled monolayers are likely about 1.0 nm thicker than the actual thickness of the monolayer. Additional errors are introduced if the value used for the escape depth is inaccurate. For example, if all escape depths are assumed to be 2.8 nm, an average thickness of 4.4 ± 0.4 nm is calculated, a difference of 20% compared to the value of 5.3 nm which was calculated above. The difficulty in establishing accurate escape depths and the problems with organic contamination result in the thickness values calculated by XPS being semi-quantitative at best. The thickness value obtained by XPS must therefore be compared to thickness values obtained by alternative methods, such as ellipsometry, to confirm the thickness of an overlayer.

The major advantage of XPS is the ability to conclusively determine the chemical structure and bonding of the organic layer. Neither ellipsometry nor SPR provide any information about the chemical characteristics of the adsorbed layer, and therefore the thickness changes measured by these techniques must assume that the layer which is adsorbing is composed of the species of interest. A second advantage of XPS is the ability to (semi-quantitatively) measure the thickness of organic layers which are adsorbed to non-conductive and non-reflective surfaces such as quartz. XPS is one of the few techniques which can be usefully applied to the study of such surfaces.

C-16 Carboxylic acid membrane. The results of the XPS studies verified the existence of a membrane consisting of the 16 carbon carboxylic acid bonded to the silicon through a dimethylsilane linkage. Inspection of the intensity values of the C(1s) peak and Si(2p) peaks of the C-16 carboxylic acid monolayer suggested that the layer was 5.3 ± 0.5 nm thick, corresponding to the presence of approximately two monolayers, assuming a thickness of 2.8 nm for a C-16 carboxylic acid monolayer.

Comparison of the thickness values obtained from XPS and ellipsometry for the C-16 carboxylic acid membranes studied show discrepancies in the overlayer thickness on the order of 100%. It is difficult to establish which of these techniques is more accurate, given that both techniques make assumptions which can produce errors in thickness of up to 50%. In addition, all ellipsometry was done on membranes which were immobilized onto silicon, while the XPS measurements were done on membranes which were immobilized onto quartz. Given the differences in the chemical composition of the two substrates, as discussed earlier, it is possible that there was a greater amount of amphiphile coated onto the quartz surfaces as compared to the silicon surfaces. This would account for the large variability in the thickness value calculated from these methods. Overall, the XPS results suggest that the carboxylic acid membrane was between one and two monolayers thick.

C-16 Carboxylic Acid Membranes with Immobilized Urease. Addition of urease to the C-16 carboxylic acid membrane resulted in the presence of a peak corresponding to nitrogen, and an increase in the C(1s) signal. This suggests that the urease was present on the membrane. The ratio of the C(1s) peak to the N(1s) peak for the urease on silicon was 4.5 while the ratio of these peaks for the C-16 carboxylic acid membrane with urease was 7.7, representing a 71% increase in the C(1s) peak owing to the presence of the membrane. The results are shown in Figs. 5a and 5b and in Table 2.

The thickness of the protein layer on the membrane was determined by calculating the ratio of the N(1s) peak of the protein to the Si(2p) peak of the substrate. This method provided a thickness value of 6.4 ± 0.6 nm. Using the total overlayer signal (C(1s) + N(1s)), the total thickness value of the membrane + urease was calculated to be 12.0 ± 1.1 nm. Subtracting the contribution from the membrane, the thickness of the urease layer was calculated to be 6.7 ± 1.2 nm, in good agreement with the thickness calculated using the N(1s) and Si(2p) signals alone. This thickness corresponds to a surface coverage of 60% of a close-packed monolayer of urease.

Urease on Silicon. The thickness value calculated from the sample containing adsorbed urease was 13.6 ± 0.9 nm, using a ratio of the N(1s) peak of the protein to the Si(2p) peak of the substrate, according to the method of Andrade.[2] This compared well with the thickness of 12.3 ± 1.6 nm which was measured using ellipsometry, and is also in agreement with the dimensions of urease. These results show that the thickness values which were calculated by XPS were accurate to within about 25% of the expected thickness value of 11 nm, and to within 11% of the thickness values which were calculated by ellipsometry.

It is interesting that the thickness values of protein layers which were obtained from XPS or ellipsometric measurements were in agreement with the thickness values expected for a monolayer of protein (in the case where a monolayer of urease was present on a silicon wafer). The use of UHV would be expected to cause substantial dehydration of the protein, and would therefore be expected to cause the protein to become compacted on the surface. This effect may also occur in cases where the sample is dried in air before ellipsometric measurements are made, but would likely be much less severe. The excellent agreement between the thickness values obtained by either XPS or ellipsometry and the expected thickness values for a monolayer of urease suggests that dehydration of the protein was not a serious problem.

Activity of Urease Bound to C-16 Carboxylic acid Membranes

We have recently developed a fluorimetric assay of enzyme activity which can be used to measure the activity of small amounts of urease immobilized onto a surface.[17] The activities of two sets of samples were tested, one which had been examined by XPS (either at one angle or at several angles) before the assay, and the second which had XPS measurements done only after the activity assays were complete. In the case where angularly resolved measurements were done on the sample, the results suggested that there was about 5.7 nm of urease on the surface, or about 50% coverage. The sample was exposed to the x-ray beam for a total of about 45 minutes, and was kept in ultra-high vacuum (UHV) for over 3 hours. The fluorimetric assay of this sample showed there to be no activity, even though the sample had been prepared only 24 hours earlier. The results suggested that exposure of the sample to the x-ray beam, in combination with exposure to ultra-high vacuum, had damaged the urease and caused complete denaturation.

In the case where XPS was done at only one angle (90°), the sample was exposed to the x-ray beam for about 10 minutes and remained in UHV for about 40

Table 3. Activity of urease immobilized onto C-16 carboxylic acid membranes as a function of time.

Day	Activity of Urease Adsorbed on Oxidized Silicon (Units.mg^{-1})	Activity of Urease immobilized on C-16 carboxylic acid membranes. Stored in buffer at 4 °C (XPS done after assay) (Units.mg^{-1}).
0	85.5 ± 7.5	‡
0.5	20.7 ± 2.1	54 ± 8
1	3.3 ± 2.0	-
2	0	33 ± 6
7	0	8.8 ± 2.4

‡ The elapsed time from preparation of the sample to the first assay was about 12 hours.

minutes. The fluorimetric assay of the sample indicated that significant activity of the enzyme remained (the activity was not quantitatively measured). These results indicate that the placement of protein coated samples into ultra-high vacuum may be done if the sample remains in vacuum for short periods of time (i.e. less than 30 mins.). The results suggest that the hydration layer of the protein is not removed immediately upon exposure to UHV. However, extended exposure to UHV combined with prolonged irradiation of the sample with x-rays does cause complete denaturation of immobilized enzymes. The results suggest that XPS measurements should be done only after the samples have been used, or that XPS should be done only at a single angle if it is done before the sample is used.

For the second set of samples, fluorimetric assays of each sample (2 samples) were done over a period of 7 days with the samples stored in phosphate buffer at 4 °C between assays. The activity values are given in Table 3. The samples were tested by XPS after all assays were complete, and indicated an average thickness of 6.7 ± 0.7 nm of urease for the samples.

The SNARF-X assay also done for samples which were composed of urease adsorbed onto silicon. The assay was done every 12 hours over a period of 2 days, after which time the samples showed no activity (see Table 3). Ellipsometry measurements suggested that the urease had not desorbed from the silicon, leaving denaturation as the likely cause of the loss of activity.

The retention of activity of urease which was immobilized onto C-16 carboxylic acid membranes, as compared to urease which was present on oxidized silicon with no spacer group, suggested that the presence of the C-16 acyl chains helped to decrease the rate of denaturation, but was unable to prevent significant denaturation over a period of a week. These results are in agreement with the results of Kallury *et al.*[1] which indicated that urease which was linked to carboxylic acid membranes with C-10 or C-23 acyl chains remained active for a period of about 7 days when stored dry. These results suggest that linkage of urease to membranes composed of ionizable phospholipids should be attempted, since these have been shown to stabilize the activity of urease to the point where 95% activity remains after a period of 7 days when stored dry, with the half-life of the activity being on the order of months.[1]

Surface Plasmon Resonance Studies of Thiol Monolayers on Gold

All SPR studies of hexadecylmercaptan (HDM) and protein adsorption used polycrystalline gold as the substrate. The use of polycrystalline gold as a medium for plasmon generation provides several advantages. The metal is relatively inert, but does provide strong binding with the sulfur group of the mercaptan for self-assembly of monolayers of high density,[18,19] and gold provides a narrow plasmon resonance linewidth which maximizes the contrast and sensitivity of detection of changes of optical mass. The purpose of this investigation was to study the dynamics of adsorption of hexadecylmercaptan onto gold and of concanavalin A (Con A) onto the mercaptan, to examine the ability of a hydrophobic membrane (surface) to immobilize a protein which is known to bind through a hydrophobic "pocket", and to establish the retention of selective chemical binding activity.

Con A is a lectin which can selectively bind to polysaccharides such as α-D-mannopyranosyl.[20-23] The protein exists as a tetramer at physiological pH, and each subunit contains a binding site for Ca^{2+} and for Mn^{2+} (co-factors which activate the saccharide binding site), a binding site for saccharide, and a hydrophobic "pocket" which permits adhesion of the protein to phospholipid membranes. The tetramer has a mass of about 100,000 Daltons, and reacts to form a precipitate with poly-saccharide due to the availability of 4 saccharide binding sites on each tetramer.

Adsorption of HDM onto clean gold surfaces was done by flowing a solution of 10 μM HDM in ethanol over the surface over a period of 12 hours, and the adsorption process was monitored by SPR. Note that all interpretations of plasmon resonance results regarding structure are treated as changes of thickness of the organic layer, owing to the sensitivity of the technique to this parameter.[10] The changes in thickness suggested that a two step (non-Langmuirian) adsorption process was occurring which was comprised of a relatively rapid initial adsorption followed by a slower step which may have involved slow adsorption of HDM onto the surface and/or rearrangements of the adsorbed membrane. SPR measurements indicated a final thickness of 3.2 $\pm$ 0.5 nm for the membrane, while ellipsometry suggested a thickness of 2.7 $\pm$ 0.4 nm. These values were in agreement with the thickness value of 2.8 nm which was calculated for a C-16 acyl chain with all-*trans* carbon-carbon bonds.

Results of experiments involving dynamic adsorption and subsequent selective binding of concanavalin A with the polysaccharides dextran and glycogen on clean gold surfaces have previously been reported,[5] and provide an interesting point of reference for the following preliminary observations involving adsorption of Con A onto HDM monolayers. The rate of adsorption of concanavalin A onto clean gold and hexadecylmercaptan surfaces was dependent on the concentration of the protein in solution. Adsorption of a compact monolayer onto clean gold occurred in 1 minute for a concentration of 80 nM of Con A, and in 60 minutes for a solution containing 4 nM of the protein. The rate was approximately 5 fold slower for adsorption of the protein onto gold that was coated with HDM. The adsorption of protein onto clean gold first occurred in a rapid process, and then slowed but continued with different kinetics after the initial period (possibly double Langmuirian fit). Such a two step process was not evident from the plasmon resonance experiments involving adsorption of Con A onto HDM coated surfaces, which indicated that adsorption of a close-packed structure of protein onto HDM occurred in a single process. The plasmon resonance experiments indicated a limiting thickness of protein of about 2 - 2.5 nm for the protein on clean gold, and a final thickness of ca. 4 nm of Con A on HDM coated surfaces (actual dimensions of monomer of protein are 4.0 x 3.9 x 4.2 nm). The substantially larger thickness obtained for the protein which was adsorbed onto HDM coated surfaces suggests that there was greater surface coverage or packing density when adsorption occurred onto the hydrophobic membrane.

For both the HDM and protein layers, the plasmon resonance experiments provided thickness estimates for the adsorbed monolayers which were somewhat larger (approx. 20 to 30 %) than the thickness values obtained by ellipsometry. The difference can likely be attributed to the removal of water and subsequent compaction of the HDM or protein layer due to drying of the samples in air for ellipsometric determinations.

The selective binding activity of the protein on the clean gold and C-16 mercaptan membrane was examined by adding glycogen or dextran to the adsorbed protein. Addition of glycogen or dextran to Con A which was adsorbed onto clean gold resulted in an increase in thickness of 0.5 - 1 nm, while addition of these species to Con A on HDM membranes resulted in a 2-fold greater increase in thickness, which is consistent with the larger amount of protein present and thus the greater number of binding sites available. In some cases, significantly larger changes in thickness (up to 5-fold) were observed upon addition of glycogen or dextran to the protein which was adsorbed onto HDM. These results show that the activity of Con A was retained, and that the the activity of the protein on the mercaptan membrane was similar to or greater than that on clean gold. The activity of the protein on HDM membranes remained unchanged for periods of at least 3 days when the samples were stored in aqueous solution.

Images which were obtained by plasmon resonance microscopy revealed that no visible structure was associated with the polysaccharide binding event on HDM membranes (even though the thickness increased significantly). These results are contrary to Fig. 6 which shows the final structure observed for the interaction of glycogen with concanavalin A that was adsorbed onto clean gold. The structural heterogeneity shown in Fig. 6 is consistent with a less ordered adsorption of concanavalin A onto the surface of clean gold. The plasmon resonance experiments indicated that the slower rate of immobilization of Con A onto mercaptan membranes was consistent with an increased density/order and a general increase of the longevity of the binding activity of the protein, while more rapid adsorption of Con A onto clean gold surfaces resulted in a non-homogeneous coverage of the surface with Con A, and an overall decrease in the binding activity of the protein.

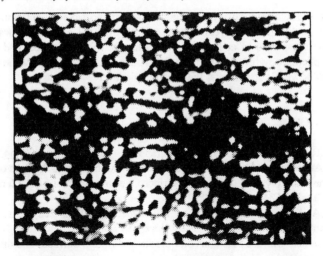

Figure 6. Surface plasmon microscopy image of the final structure observed for polysaccharide interaction with concanavalin A that was adsorbed to clean gold. Image field is 400 μm wide.

4 CONCLUSIONS

Long-chain carboxylic acids were immobilized by silane coupling to quartz or silicon substrates, while long-chain thiols were immobilized onto metallic gold substrates by self-assembly. Active proteins were subsequently attached to both of these coatings. Ellipsometry and XPS were used to determine the structure of the silane-based films on quartz and silicon, respectively, and the extent of coverage of substrates with amphiphiles and proteins. The thickness values reported by XPS for the C-16 carboxylic acid membrane were about 2-fold greater than those measured by ellipsometry. The discrepancy in the thickness values likely originated from the differences in the hydroxyl density at the surface of quartz and silicon wafers. XPS and ellipsometry provided similar thickness values for urease layers which were immobilized onto the hexadecanoic acid membrane. The thickness corresponded to about 50% to 60% coverage of the protein. All thickness values reported by XPS and ellipsometry were relative values as it was not possible to accurately establish the absolute thickness of organic overlayers using either technique. It was shown that XPS does not significantly damage protein samples if measurements are made at only a single angle, though use of angularly resolved measurements caused total denaturation of the protein, owing to dehydration of the protein and sample damage which was caused by prolonged exposure to x-ray radiation.

Surface plasmon resonance microscopy was used to investigate the *in situ* dynamic properties of the thiol-based films on gold, as well as the kinetics of protein adsorption to these films. It was shown that the SPR technique could be used to follow the dynamic adsorption of either an amphiphilic monolayer onto gold, or a protein layer onto the surface of a hexadecylmercaptan membrane. The thickness values reported by SPR were relative values since the difficulty in establishing the refractive index of the adsorbed layers made it impossible to determine absolute thickness values for these layers. The use of SPR did not cause denaturation of the protein over a period of at least 3 days. Images which were obtained by surface plasmon microscopy revealed that the Con A adsorbed as a compact, homogeneous layer with no visible structure at the microscopic level.

REFERENCES

1. K.M.R. Kallury, W.E. Lee and M. Thompson, Anal. Chem., 1992, 64, 1145.
2. J.D. Andrade, in J.D. Andrade (Ed.) 'Surface and Interfacial Aspects of Biomedical Polymers, Volume 1, Surface Chemistry and Physics', Plenum Press, New York, 1985, pp. 105-195.
3. R.M.A. Azzam and N.M. Bashara, 'Ellipsometry and Polarized Light', North-Holland Publishing Company, New York, 1977.
4. H. Raether, 'Surface Plasmons on Smooth and Rough Surfaces and on Gratings', Springer, Berlin, 1988.
5. R.F. De Bono, U.J. Krull and Gh. Rounaghi, in P.R. Mathewson and J.W. Finley, (Eds.) 'Biosensor Design and Application', ACS symposium series 511, American Chemical Society, Washington, DC, 1992, pp. 121-136.
6. V. Tscharner and H.M. McConnell, Biophys. J., 1981, 36, 409.
7. M. Losche, E. Sackmann and H. Mohwald, Ber. Bunsen-Ges. Phys. Chem., 1983, 87, 848.
8. B. Rothenhausler and W. Knoll, Nature, 1988, 332, 615.
9. W. Hickel and W. Knoll, J. Appl. Phys., 1990, 67, 3572.
10. R.P.H. Kooyman and U.J. Krull, Langmuir, 1991, 7, 1506.
11. F.L. McCracken, 'NBS Technical Note' 479, Washington DC, 1969.
12. C.P. Tripp and M.L. Hair, Appl. Spectrosc., 1992, 46, 100.
13. D. Ducharme, J.J. Max, C. Salesse and R.M. Leblanc, J. Phys. Chem., 1990, 94, 1925.
14. J. Gun, R. Iscovici and J. Sagiv, J. Coll. Int. Sci., 1984, 101, 201.
15. C.R. Cantor and P.R. Schimmel, 'Biophysical Chemistry', Vol. 2, Freeman, San Francisco, 1980, p. 554.
16. R. Martinez-Zaguilan, G. M. Martinez, F. Lattanzio and R.J. Gillies, Am. J. Physiol., 1991, 260, C297.
17. J.D. Brennan, V. Kukavica, K.M.R. Kallury and U.J. Krull, Can. J. Chem., submitted.
18. M.D. Porter, T.B. Bright, D.L. Allara and C.E.D. Chidsey, J. Am. Chem. Soc., 1987, 109, 3559.
19. C.D. Bain, E.B. Troughton, Y-T. Tao, J. Evall, G.M. Whitesides and R.G. Nuzzo, J. Am. Chem. Soc., 1989, 111, 321.
20. L.L. So and I.J. Goldstein, J. Biol. Chem., 1967, 242, 1617.
21. J.L. Ochoa, T. Kristiansen and S. Pahlman, Biochim. Biophys. Acta, 1979, 577, 102.
22. G.M. Edelman and J.L. Wang, J. Biol. Chem., 1978, 353, 3016.
23. G.N. Reeker Jr., J.W. Beker and G.M. Edelman, J. Biol. Chem., 1975, 250, 1525.

Spectroscopy of Evaporated and Langmuir–Blodgett Films of Gadolinium Bisphthalocyanine on Metal Surfaces

B. Berno, R. Aroca and A. Nazri

MATERIALS AND SURFACE SCIENCE GROUP, DEPARTMENT OF
CHEMISTRY AND BIOCHEMISTRY, UNIVERSITY OF WINDSOR,
WINDSOR, ONTARIO N9B 3P4, CANADA

1 INTRODUCTION

Phthalocyanine dyes are a very special class of robust organic materials that have found a wide range of applications in thin film devices.[1-3] Recently, they have also been studied for nonlinear optical applications.[4] Gadolinium (III) bisphthalocyanine is an important member of the lanthanide series of compounds where the central metal ion Gd^{3+} has a $4f^7$ electron configuration and an ionic radius of 93.8 pm, the middle point in the ion contraction observed in the Ln series. The synthesis of $GdPc_2$ was initially reported by Kirin et al.[5] A $GdHPc_2$ molecular formula was assumed by M'Sadak et al.[6] for the material synthesized, where Pc denotes the $[C_{32}H_{16}N_8]^{2-}$ ligand. The NMR spectra and the symmetry of the $[LnPc_2]^-$ anions, including $[GdPc_2]^-$, have also been reported.[7] It was found that $[LnPc_2]^-$ anion in solution could be assigned to the D_{4d} point group symmetry where the Pc rings take a staggered configuration. The reduced form of $GdPc_2$ was found to belong to a subgroup of the D_{4d} group. The electronic absorption spectra of $LnPc_2$ implied the existence of an exciton interaction between the two chromophores as had been previously suggested.[8] MacKay et al.[9] discussed their preparation of two gadolinium bisphthalocyanines which produced different coloured solutions according to the solvent: green in benzene and blue in dimethylformamide. Both of these forms showed a well defined EPR signal.[9] Infrared bands in the fingerprint region were also listed in reference [9]. The Raman spectra and the vibrational interpretation for $GdPc_2$ frequencies have not yet been reported. In the present work a spectroscopic characterization of $GdPc_2$ thin solid films evaporated on several substrates as well as the fabrication of LB films and their Raman spectra are reported.

2 EXPERIMENTAL

Silver films for reflection-absorption infrared spectroscopy (RAIRS) were prepared by evaporation of Ag metal to a thickness of 110 nm onto a glass substrate held at 200 °C. The deposition rate was held between 3-5 Å/s.

The preparation of Au island substrates for SERS was similar to the Ag film preparation except the rate of deposition was held at 0.5 Å/s and the final thickness was approximately 40 Å. Thin films of $GdPc_2$ were evaporated onto KBr pellets, glass slides (Corning 7059), Ag thin films, Au island films, and grids for transmission electron microscopy (TEM). The organolanthanide thin films were deposited at a rate of 2 Å/s onto substrates at ambient temperature. The thicknesses for all evaporated films were monitored by XTC Inficon crystal oscillators.

Langmuir monolayers were prepared in a Lauda Langmuir film balance. A toluene solution was spread onto the aqueous subphase maintained at 15 °C. Monolayer deposition was carried out by a Lauda Filmlift FL-1 electronically controlled dipping device. Monolayers were transferred at 2 mm/min at a pressure of 15 mN/m.

Infrared spectra were recorded on a Bruker FTIR Model 98 spectrometer operated under vacuum. RAIR spectra were recorded with the aid of a Spectra Tech Inc. Specular Refl. Model #500 arranged at 75° to the incident light. Spectral resolutions were 1 cm^{-1} and 2 cm^{-1} for transmission and RAIR experiments respectively.

Scattering experiments were performed using a Spectra Physics Model 2020 Kr^+ laser operating at 647.1 nm as the light source. The Stokes inelastic scattering was measured with a Spex-1403 double spectrometer.

Electronic transitions were observed with a Perkin-Elmer Lambda 9 UV/VIS/NIR spectrometer interfaced to a Perkin-Elmer PE-7700 series computer.

3 RESULTS AND DISCUSSION

<u>Sample Characterization</u>

The sample of $GdPc_2$ was kindly provided by Dr. L.G. Tomilova.[10] The EPR spectra of the solid confirmed the free radical nature of the $GdPc_2$ material under study. The absorption spectrum of the material in chloroform corresponds to the green gadolinium bisphthalocyanine spectrum given by MacKay et al.[9] with Q band maximum at 670 nm. For the material dissolved in toluene the maximum was measured at 667 nm and a shoulder at 639 was evident that could correspond to the blue form.[9] The relative intensities in the electronic absorption spectrum were clearly affected by solvent interactions and the green form was stabilized by the polar solvent. The near infrared spectrum of $GdPc_2$ in the 1000-2000 nm range was identical with that recently reported by Shirk et al.[4]

The transmission electron microscopy of an evaporated film 200 nm thick is shown in Figure 1. The aggregation and formation of microcrystals are evident in the micrograph of the film. The molecular organization in these structures will be discussed later.

The Langmuir layer was formed on the water surface and the surface-pressure area isotherm (π-A) was recorded in two separate sets of experiments. The π-A isotherm showed a low compressibility region at low pressures (0-15 mN/m) with a limiting area of 0.60 nm^2, and a typical solid

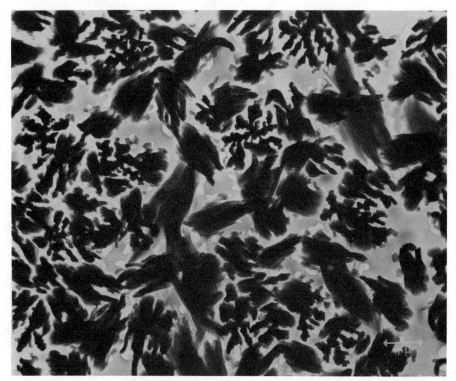

Figure 1 Transmission electron microscopy of a 25 nm film of GdPc$_2$ with 8 x 10^3 magnification.

phase low compressibility region between 15 and 50 mN/m. The limiting area for the latter part of the curve was about 0.40 nm^2. The measurements indicate that solid phase does represent a single monolayer, and either a folding (bilayer formation) occurs or aggregates are formed on the surface before and during the compression of the monolayer. The organization in Langmuir-Blodgett (LB) films fabricated from the floating layers will be discussed with the information of the infrared data.

<u>Resonant Raman and SERRS of Thin Solid Films</u>

The absorption spectrum shows that the 647.1 nm laser line of the Kr$^+$ ion laser is in resonance with the Q band of GdPc$_2$. The resonant Raman scattering (RRS) of the material in a KBr pellet and of thin solid film (200 nm) on Kbr were recorded. The SERRS of a casted film on a Au island film and a LB monolayer on Au were also recorded. All the RRS and SERRS spectra were identical, and as an example the SERRS spectrum of one LB monolayer on Au is given in Figure 2. Clearly the GdPc$_2$ films were not affected by the substrate. Hence, the monolayer was physisorbed onto the Au island film. The RRS spectrum of GdPc$_2$ is typical of most LnPc$_2$ complexes with

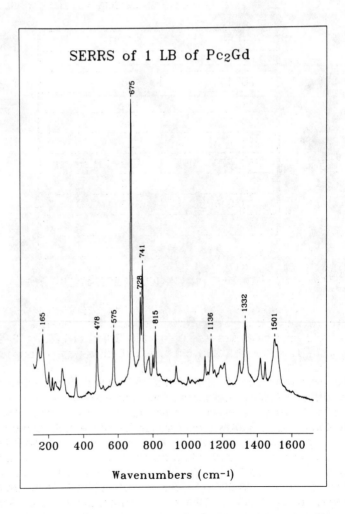

Figure 2 SERRS of a single LB monolayer of GdPc$_2$ deposited on Au island films.

characteristic frequencies of the macrocycle at 675 cm^{-1}, 741 cm^{-1}, 1332 cm^{-1} and 1501 cm^{-1}. The observation of the 728 cm^{-1} band in the spectrum is an indication of the presence of certain amount of blue form in the sample.[11] A comparison of the observed infrared and Raman frequencies given in Tables 1 and 2 excludes the D_{4d} point group symmetry for the molecule. The exact point group can not be determined from the data set, and it could be any of the subgroups of the correlation table for descent in D_{4d} symmetry.

Table 1. Observed far infrared and corresponding Raman frequencies (cm^{-1}).

FAR-IR	SERRS
106	104
120, 143	138
171	165
190, 202	202
237	222
259, 276	279 M-N stretch
307	290
343, 346, 381	352
425	427
	477

Transmission and Reflection-Absorption Infrared Spectra

The infrared spectrum of GdPc$_2$ material dispersed in a Kbr matrix is given in Figure 3. It can be seen that the four strong infrared bands seen in most LnPc$_2$ molecules are also present in GdPc$_2$ at 725, 1113, 1318 and 1444 cm^{-1}. However, the 1524 cm^{-1} band which is weak or absent in other LnPc$_2$ molecules (but Raman active) is seen with a strong relative intensity in the gadolinium complex. It could be concluded that the macrocycle distortion in the GdPc$_2$ molecule is considerable when compare with LuPc$_2$[12] and YPc$_2$[13] which results in a lower molecular symmetry. The spectra presented in Figure 3 allow a brief discussion of the molecular organization in the evaporated film. The reference point for the discussion is the out-of-plane C-H vibration which is observed at 725 cm^{-1}. The transition dipole moment for this particular vibration is perpendicular to the Pc plane (assuming the macrocycles are planar or quasi-planar). In the transmission geometry, the electric field of the incident light is parallel to the film surface and minimum intensity would be observed for the C-H wag when the macrocycle plane is on a face-on molecular orientation. The relative intensity of this band, as can be seen in Figure 3, has diminished if compared with the 764, 1364 and 1524 cm^{-1} bands in the spectrum of the pellet. Since the last three frequencies can be assigned to in-plane vibrations of the macrocycle, it can be concluded that there exist a prevalent tilted molecular organization in the evaporated film. According to the electron microscopy data, the organization exists in the microcrystals formed on the KBr surface, probably due to molecular stacking of the Pc rings. It should be pointed out that the C-H wagging would reach maximum relative intensity (dipole-field interaction) in an anisotropic film with perfect edge-on molecular orientation. Since the stacking takes place in all directions in the **x-y** (surface plane), the intensity of the wagging vibration would also be diminished.

The transmission infrared spectrum of 18 LB layers on ZnS was again

Table 2:   Observed infrared and Raman frequencies (in cm⁻¹).

IR pellet	IR film	RAIRS	SERRS	Assignment
557 (w)	556 (w)	557 (w)		Benz. radial
			575 (m)	
640 (w)	640 (w)	640 (w)		
678 (vw)			676 (vs)	Pc breathing
725 (s)	726 (m)		729 (m)	C-H wag
740 (m)	743 (w)		740 (s)	Pc ring
767 (m)	764 (m)	764 (s)		Pc ring
778 (w)			779 (w)	Pc ring
			800 (w)	
810 (vw)			814 (m)	Pc ring
839 (vw)	839 (vw)	839 (vw)		
857 (vw)	857 (vw)	858 (vw)		
882 (w)	883 (vw)			
943 (vw)			935 (w)	
1001 (vw)			1004 (w)	
	1080 (br)			
1113 (m)	1113 (w)		1103 (w)	
			1137 (w)	
1160 (w)		1168 (vw)	1156 (vw)	
1190 (vw)	1188 (vw)	1190 (vw)	1190 (vw)	
1318 (s)	1295 (vw)	1295 (vw)	1300 (w)	C-H bend
	1331 (vw)		1332 (m)	Pyrrole st.
			1337 (m)	Pyrrole st.
1362 (s)	1363 (m)	1364 (m)		
	1399 (w)	1392 (vw)	1420 (w)	Isoindole st.
1444 (m)		1443 (w)	1444 (w)	Isoindole st.
1500 (sh)			1500 (w)	Pyrrole st.
1524 (s)	1524 (s)	1527 (s)	1515 (w)	Aza stretch
1579 (w), 1584 (w)	1579 (w)	1580 (w)		Benz. stretch
1606 (w), 1637 (vw)		1635 (vw)		Benz. stretch

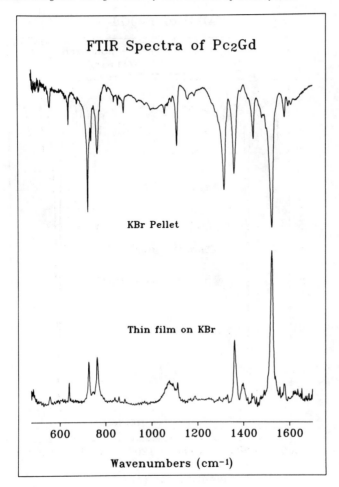

Figure 3 FTIR spectra of a KBr pellet of GdPc$_2$ and of a 25 nm evaporated film on a KBr surface.

different from that of the evaporated film with the 725 cm^{-1} band being the strongest in the spectrum. At the same time the set of bands at 764, 1364 and 1524 cm^{-1} showed a decrease in relative intensity. The results indicate that in the multilayer assembly there is a more anisotropic organization of edge-on (macrocycle edge) molecular orientation.

The reflection-absorption spectra of GdPc$_2$, 25 nm film on Ag, are presented in Figure 4. In the RAIRS experiments, no detectable intensity was obtained for incident radiation polarized parallel to the film surface (S-polarization in Figure 4). In the RAIRS geometry only the E$_z$ (P-polarization)

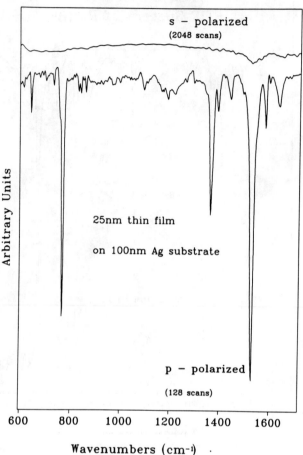

Figure 4 Reflection-absorption FTIR spectra of a 25 nm film of GdPc$_2$ on Ag: 2048 scans of the S-polarized incident light; 128 scans of the P-polarized incident light.

component of the incident light can interact with the adsorbate, and the surface selection rules lead to the manifestation of molecular vibrations with a finite component of their dynamic dipole perpendicular to the metal surface.[14] The IR bands that were intense in the spectrum of the 200 nm film (764, 1364 and 1524 cm^{-1} bands) were again enhanced in the RAIR spectrum. Since dynamic dipole components perpendicular to the metal surface enhance their interaction with the external field, every molecule of a tilted molecular organization could contribute to the relative intensity of the

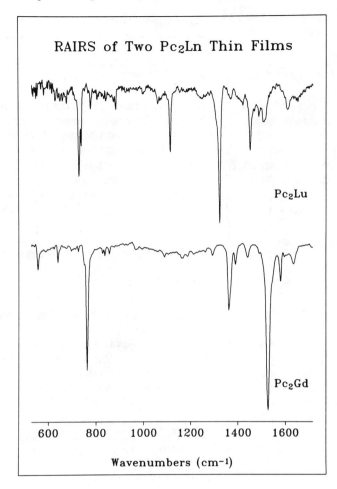

Figure 5 Comparative RAIR spectra of similar 25 nm films of LuPc$_2$ and GdPc$_2$ evaporated onto Ag.

in-plane (macrocycle plane) modes. The RAIR spectra also support a tilted organization of the Pc molecules on the metal surface.

It is important to notice that the packing arrangement in LnPc$_2$ complexes may have a pattern or could be different depending on the central cation. To illustrate this point, the RAIR spectra of a 25 nm film of LuPc$_2$ and GdPc$_2$, prepared under identical conditions are shown in Figure 5. It can be seen that a different set of fundamental frequencies are enhanced on the metal surface, a clear indication of a different molecular orientation in each Pc packing of the evaporated film. The pattern in film morphology for evaporated films and LB film of LnPc$_2$ complexes will be the natural extension of the present work.

4 CONCLUSIONS

Thin solid films of GdPc$_2$ have been prepared and studied using optical spectroscopy. Evaporated films were found to be formed by microcrystals. However, molecular orientation studies using infrared data showed that there exist a prevailing tilted molecular orientation in the film. LB film fabricated on ZnS had a very clear edge-on molecular orientation according to FTIR spectra. The molecular organization in thin solid films of LnPc$_2$ represent an interesting case study for the relationship between the central metal atom and the packing in the evaporated and LB films.

ACKNOWLEDGEMENTS
Financial support from NSERC Canada is gratefully acknowledged.

REFERENCES

1. A.L. Thomas, "Phthalocyanines Research and Applications", CRC Press, Inc., Boca Raton, Florida, 1990.
2. "Phthalocyanines Properties and Applications", C.C. Leznoff, A.B.P. Lever, Eds., VCH Publishers, Inc., New York, 1989.
3. J. Simon, J.J. Andre, "Molecular Semiconductors", J.M. Lenh, C.W. Rees, Eds., Springer-Verlag, Berlin, 1985.
4. J.S. Shirk, J.R. Lindle, F.J. Bartoli, M.E. Boyle, J. Phys. Chem., 1992, 96, 5847.
5. I.S Kirin, P.N. Moskalev, Y.A. Makashev, Russ. J. Inorg. Chem, 1967, 12, 369.
6. M. M'Sadac, J. Roncali, F. Garnier, J. Electroanal. Chem., 1985, 189, 99.
7. H. Konami, M. Hatano, A. Tajiri, Chem. Phys. Lett., 1989, 160, 163.
8. M. Moussavi, A. De Cian, J. Fisher, R. Weiss, Inorg. Chem., 1988, 27, 1287.
9. A.G. MacKay, J.F. Boas, G.J. Troup, Aust. J. Chem., 1974, 27, 955.
10. L.G. Tomilova, E.V. Chernykh, N.T. Ioffe, E.A. Luk'yanets, Zh. Obshch. Khim., 1984, 53, 2339.
11. D. Battisti, L.G. Tomilova, R. Aroca, Chem. of Materials, 1992, 4, 1323.
12. R. Aroca, R.E. Clavijo, C.A. Jennings, G.J. Kovacs, J.M.Duff and R.O. Loutfy, Spectrochim. Acta, 1989, 45A, 957.
13. M.L. Rodriguez-Mendez, R. Aroca, J.A. DeSaja, Spectrochim. Acta, 1993, 49A, 965.
14. B.E. Hayden, in Vibrational Spectroscopy of Molecules on Surfaces, "Methods of Surface Characterization", Vol. 1., J.T. Yates, T.E. Madey, Eds., Plenum Press, New York, 1987, 267.

Surface Chemistry of Microporous Manganese Oxides

S. L. Suib, H. Cao, and W. S. Willis

DEPARTMENTS OF CHEMISTRY AND CHEMICAL ENGINEERING,
U-60, UNIVERSITY OF CONNECTICUT, STORRS, CT 06269-3060, USA

1   INTRODUCTION

Fundamental knowledge about surfaces of microporous manganese oxide materials such as factors controlling mixed valency, migration of atoms during photolysis or thermal treatment, precise binding energies, the nature of adsorption sites and similar phenomena are not well understood. With oxide materials, it is well known that oxide vacancies exist and can be induced by chemical methods.[1] In addition, several types of surface oxygen species have been identified on metal oxide surfaces such as adsorbed molecular oxygen, adsorbed atomic oxygen, and oxygen that can be desorbed under high thermal conditions that is attributed to lattice oxygen.[2] How such species are evolved and participate in chemical reactions is continually under investigation.

The catalytic activity of metal oxides in reactions such as isomerization,[3] hydrogenation,[4] and selective oxidations[5] is often related to several of the phenomena described above. Surface acidity, the degree of surface coordinative unsaturation, and specific phenomena such as decoration and surface reduction play an important role in the chemistry of such systems.[1] Considerable attention has been placed in recent years on the preparation and characterization of metal oxides that are used in these catalytic reactions.

There has been considerably less research in the area of photochemical activation of metal oxide systems for the purpose of photoassisted catalytic transformations.[6] Some of the reactions that have been pursued include hydrocarbon oxidations,[7] oxidation of alcohols,[8] oxidation of small molecules[9] and exchange reactions.[10] Much of this research has

centered around the use of titania, either in the
rutile or anatase structures.[11] Both liquid and gas
phase reactions have been studied.

The purpose of this paper is to discuss some
recent results we have obtained for photoassisted
catalytic oxidation reactions of isopropanol over a
variety of manganese oxide catalysts. These gas phase
reactions have been studied over both crystalline and
amorphous manganese oxide systems. Such materials
have been characterized by a variety of methods
including X-ray powder diffraction, X-ray
photoelectron spectroscopy, surface area, temperature
programmed desorption and others.

2   EXPERIMENTAL SECTION

This section describes the preparation of manganese
oxide materials, their characterization and the
photoassisted catalytic oxidation reactions.

Preparation of Managanese Oxides

Crystalline manganese oxide materials that have
a tunnel structure were prepared by literature
methods. These materials include synthetic todorokite
known as octahedral molecular sieve 1 (OMS-1)[12] and
synthetic cryptomelane[13] (OMS-2). The structures of
these materials are shown in Figure 1.

(3 x 3)   **(2 x 2)**

Figure 1. Structure of OMS-1 (3 x 3) and
           OMS-2 (2 x 2).

The OMS-1 structure consists of 3 vertex and
edge shared MnO_6 octahedra on each side (3 x 3)
having a 6.9 Å pore size whereas the OMS-2 structure
consists of a (2 x 2) structure of 4.6 Å pore size.

Amorphous manganese oxide catalysts were
prepared by mixing potassium permanganate with oxalic
acid or manganese (II) acetate. In a typical
preparation, 1.58 g $KMnO_4$ was dissolved in 60 mL
distilled deionized water and mixed with 2.28 g

oxalic acid dissolved in 100 mL distilled deionized water. Precipitates were washed with distilled deionized water and then vacuum dried prior to use in photoassisted catalytic oxidation reactions. The objective in these preparations was to prepare a mixed valent manganese oxide series of materials where the oxidation state could be tailored by adding different ratios of $KMnO_4$ and oxalic acid.

Characterization Studies.

All catalysts were characterized with X-ray powder diffraction (XRD), X-ray photoelectron spectroscopy (XPS), surface area, energy dispersive X-ray (EDX) analyses, differential scanning calorimetry (DSC), thermogravimetric analysis (TGA) and Fourier transform infrared (FTIR) spectroscopy. Details of these types of experiments can be found in the literature.[14-16]

Photoassisted Catalytic Oxidations

Photoassisted catalytic oxidations of isopropanol were done in a flow reactor with 0.025 g catalyst which was placed on a stainless steel grid inside a sealed stainless steel reactor which had a window to allow in visble light from a 1000 W Xe lamp. The radiation from the lamp was filtered first through a water bottle to eliminate infrared radiation and then through a 425 nm cutoff filter to eliminate ultraviolet radiation. Lines in and out of the reactor from the isopropanol bubbler were heated with heating tape to about 75°C to avoid condensation in the lines. Different carrier gases such as N_2 and O_2 were used to transport isopropanol into the reactor. Products were collected in a dry ice acetone cold trap and syringed into a gas chromatograph with a thermoconductivity detector for analysis. Columns used for separations were 10% Carbowax 20 M on 60-80 mesh Anakrom on C-22 firebrick. Helium carrier gas at a flow rate of 30 mL/min was used. The detector was set at 250°C, the injector port at 250°C and the oven temperature was at 100°C for these isothermal separations.

3   RESULTS

Catalyst Synthesis and Characterization

OMS-1 in the Mg ion-exchange form and OMS-2 in the K^+ form were successfully prepared on the basis of XRD, elemental analysis, EPR, surface area and thermal analyses. All materials prepared by the addition of $KMnO_4$ to oxalic acid resulted in amorphous materials on the basis of XRD experiments. Temperature programmed desorption, thermogravimetric analyses, differential scanning calorimetry and FTIR experiments for the amorphous MnO_2-2-2 catalyst show that H_2O and oxalic acid ligands are lost at temperatures up to 100°C, leaving essentially a manganese oxide amorphous material. At 130°C, MnO_2-2-2 loses oxygen, whereas commercial manganese oxide materials lose oxygen at considerably higher temperatures with bulk oxygen being evolved near 550°C. OMS-1 loses oxygen in at least two stages up to 600°C and OMS-2 loses oxygen at even higher temperatures, bulk oxygen being released up to 700-800°C.

In situ XPS data for the photolyzed amorphous MnO_2-2-2 catalyst with respect to the unphotolyzed MnO_2-2-2 material show an increase in O/Mn ratio of 5.5. Such a change in surface oxygen during photolysis did not occur for commercial manganese oxide, OMS-1 or OMS-2. Temperature programmed desorption studies of OMS-1 and OMS-2 with He carrier gas show a loss of oxygen in three different temperature regimes. For OMS-1, these ranges include 200-300°C, 300-475°C and 475-600°C. Similar data are obtained for OMS-2; however, these three ranges are at higher temperatures, the latter being between 600-800°C. TGA data for MnO_2-2-2 show a major weight loss at 300°C, which is considerably lower than either OMS-1 (600°C), OMS-2 (800°C) or commercial manganese oxide (600°C).

Photoassisted Catalytic Oxidation of Isopropanol

Conversions of the photoassisted catalytic oxidation of isopropanol to acetone with OMS-1, OMS-2 or amorphous MnO_2-2-2 were always enhanced when O_2 carrier gas was used in comparison to N_2 carrier gas. Prephotolyzing all three catalyst systems in the absence of alcohol substrate led to a decrease in overall conversion after isopropanol was later added. FTIR data of the materials that were shown to be active showed absorptions in the 3600-3100 cm^{-1} range.

Thoroughly dehydrated $MnSO_4$ systems were not active at all and showed no IR bands between 3600-3100 cm^{-1}. The surface area of MnO_2-2-2 was about 164 m^2/g and OMS-1 had about a 160 m^2/g surface area. A comparison of the turnover rates of formation of acetone are summarized in Table 1. In the absence of manganese oxide catalysts there was no conversion of isopropanol to acetone. As power was increased from 200 W to 950 W the rate of formation of acetone linearly increased with all catalyst systems. There was no loss of crystallinity of either OMS-1 or OMS-2 during photolysis experiments on the basis of XRD experiments of the catalysts after photolysis.

In all of the above photolysis experiments the only product observed was acetone from gas chromatographic analyses. Some longterm experiments were carried out by taking a catalyst that was used for up to 10 h and then using it again under the same initial conditions as a fresh catalyst. The rate of formation of acetone for the used MnO_2-2-2 catalysts was the same as the fresh MnO_2-2-2 catalyst at least initially (first 5-7 h). Electron microscopy data taken before and after photolysis show no differences in morphologies of these materials. OMS-1 and OMS-2 have platelike and fibrous morphologies, respectively

Table 1. Comparison of Turnover Rates for Different Catalysts.

Catalyst	Turnover Rate[*]
OMS-1	1.1×10^{-3}
OMS-2	1.1×10^{-4}
MnO_2-2-2	1.7×10^{-3}

[*] - mol acetone/h-g catalyst

whereas MnO_2-2-2 is globular with no apparent crystalline shape.

4   DISCUSSION

Synthesis and Characterization

Synthesis and characterization results show that OMS-1 and OMS-2 have been prepared that are stable under the conditions used for photoassisted catalytic oxidation of isopropanol. The MnO_2-2-2 catalyst has been shown to be amorphous and moderately thermally stable with respect to other manganese oxide systems. The surface areas of all materials are roughly the same.

The exact nature of MnO_2-2-2 materials is not well understood. FTIR data show that there is very little oxalate ligand left after thermal treatment to $100^{\circ}C$ in vacuum of the amorphous gel precursor. The globular shape from electron microscopy data is in line with an amorphous material in contrast to the platelike and fibrous morphologies of OMS-1 and OMS-2, respectively.

Photoassisted Catalytic Oxidation of Isopropanol

Photoassisted catalytic oxidations of isopropanol with manganese oxide catalysts have considerably higher conversion in the presence of oxygen than in nitrogen. These data suggest that manganese ions may be reduced during photooxidation and regeneration of higher valent manganese oxidation states is necessary by using oxygen in the feed stream. All of the manganese oxide materials (OMS-1, OMS-2 and MnO_2-2-2) are more active than uranyl supported zeolite, clay or pillared clay catalysts[17] that have previously been studied. The selectivity to acetone with all these materials is 100%.

The turnover rates of Table 1 clearly show that the amorphous MnO_2-2-2 catalyst is more active than the OMS materials. The surface areas of these three catalysts are similar so the activity differences do not appear to be due to differences in surface area. All materials are black and it is unlikely that differences in absorption cross sections for these three catalysts can cause a change in activity. FTIR data for active catalysts show a good correlation with the presence of surface hydroxyl groups in these materials. When surface hydroxyl groups are not present, conversion either decreases (dehydration of an active catalyst) or is zero (non-hydroxyl containing catalysts like $MnSO_4$).

The differences in turnover rates of formation of acetone for the three different catalysts are very likely related to differences in surface functional groups and the behavior of such groups during photolysis. A discussion of these differences is given below.

Surface Chemistry of Manganese Oxide

The most active catalyst MnO_2-2-2 is the least thermally stable material. It evolves oxygen species at temperatures near 300°C whereas OMS-1 and OMS-2 evolve similar species at much higher temperatures (600-800°C). Hydroxyl groups on the surface of all three catalysts also seem to play a role in the photoassisted catalytic oxidation, since the absence of such groups leads to a total loss in activity. These data suggest that oxygen transfer alone does not seem to be important in these materials, for example loss of lattice oxygen alone is not sufficient.[1]

The in situ XPS data clearly show that MnO_2-2-2 during photolysis shows an enhancement of surface oxygen. This suggests that lattice oxygen is concentrated at the surface of MnO_2-2-2 to a greater extent during photolysis than OMS-1 or OMS-2. An interesting trend is that the lower the temperature of oxygen loss, the higher the conversion. It is not clear whether oxygen migrates to the surface during photolysis or whether manganese migrates into the interior of MnO_2-2-2 or whether both processes occur. There does not appear to be any major loss of manganese from the solid during photolysis.

5   CONCLUSIONS

We have shown here that amorphous manganese oxide catalysts are more active in the conversion of isopropanol to acetone than crystalline manganese oxide materials. Increased conversion appears to be related to the amount of surface oxygen and lower temperatures of desorption of oxygen from such systems. All materials are selective catalysts. The presence of hydroxyl groups as detected by FTIR experiments seem to be important in such photolyses. Mechanistic data need to be collected in order to further understand the exact nature of oxygen in such systems and how further optimization of this process might be accomplished. Experiments with labeled oxygen on the surface of the manganese oxide materials and incorporated into the reactant isopropanol as well as deuterium labeling experiments are currently underway in order to answer such questions.

6   ACKNOWLEDGMENTS

We thank the Department of Energy, Office of Basic Energy Sciences, Divsion of Chemical Sciences for support of this research.

7   REFERENCES

1. H. H. Kung, <u>Transition Metal Oxides: Surface Chemistry</u> and <u>Catalysis</u>, Elsevier, Amsterdam, 1989.
2. M. Iwamoto, Y. Yoda, N. Tamazoe, T. Seiyama, <u>J. Phys. Chem.</u>, 1978, <u>82</u>, 2564.
3. C. Y. Hsu, C. R. Heimbuch, C. T. Armes, B. C. Gates, <u>J. Chem., Soc. Chem. Comm.</u>, 1992, 1645-1646.
4. H. C. Foley, A. W. Wang, B. Johnson, J. N. Armor, in <u>Selectivity in Catalysis</u>, M. E. Davis, S. L. Suib,   Eds., ACS Symposium Series, 517, ACS, Washington, DC, 1993, 168.

5. R. A. Periana, D. J. Taube, E. R. Evitt, D. G.
   Loffler, P. R. Wentrcek, G. Voss, T. Masuda
   Science, 1993, 259, 340-343.
6. S. L. Suib, Chem. Rev., 1993, 93, 803.
7. G. M. Bancroft, A. M. Draper, M. M. Hyland, P.
   Demayo, New J. Chem., 1990, 14, 5.
8. R. Kunneth, C. Feldmer, H. Kisch,
   Ang. Chem. Int. Ed. Engl., 1992, 31, 1039.
9. N. W. Cant, J. R. Cole, J. Catal., 1992, 134,
   317.
10. J. Cunningham, E. Goold, J. Chem. Soc. Far.Trans.
    1, 1982, 78, 785.
11. J. F. Tanguay, S. L. Suib, R. W. Coughlin,
    J.Catal.,1989, 117, 335.
12. Y. F. Shen, R. P. Zerger, R. DeGuzman, S. L. Suib,
    L. McCurdy, D. I. Potter, C. L. O'Young, Science, 1993,
    260, 511.
13. R. DeGuzman, Y. F. Shen, B. R. Shaw, S. L. Suib,
    C. L. O'Young, manuscript in preparation.
14. M. A. Kmetz, J. L. Laliberte, S. L. Suib, F. S. Galasso.
    Ceram. Eng. Sci. Proc. 1992, 13, 743.
15. G. J. Colpas, M. J. Maroney, C. Bagyinka, M. Kumar,
    W. S. Willis, S. L. Suib, P. K. Mascharak,
    N. Baidya, Inorg. Chem., 1991, 30, 920.
16. R. K. Force, R. P. Grosso, S. McClain, W. S.
    Willis, S. L. Suib, Inorg. Chem., 1990, 29, 1924.
17. S. L. Suib, J. F. Tanguay, M. L. Occelli,
    J. Am. Chem. Soc., 1986, 108, 6972.

5 key words

manganese oxides, heterogeneous photocatalysis,
surface migration, isopropanol oxidation, octahedral
molecular sieves.

High Temperature Sorbents for Oxygen Supported on Platinum Modified Zeolites

Pramod K. Sharma

JET PROPULSION LABORATORY, CALIFORNIA INSTITUTE OF
TECHNOLOGY, PASADENA, CA 91109, USA

1 ABSTRACT

The role of platinum in the reduction of metals like cobalt and copper when added to a zeolite by ion exchange is investigated. It is seen that the platinum addition has a favorable effect in both the reduction process as well as in subsequent oxygen uptake. Transmission electron microscopy of the reduced samples of copper exchanged zeolites indicates that smaller copper crystallites are obtained when platinum is present. The higher reactivity of the reduced sorbent results, at least in part, due to the smaller and better dispersed crystallites.

2 INTRODUCTION

Copper exchanged Linde zeolites 13X and L upon reduction are shown[1-3] to result in efficient sorbents for oxygen. Copper utilization in excess of 50 percent can be achieved while equilibrium oxygen levels can be maintained below 1 ppb (parts-per-billion) at temperatures up to 600 °C. However, special materials processing applications require oxygen removal to parts-per-trillion (ppt) levels at temperatures up to 1000 °C. For example, processing of certain materials in the semiconductor industry and processing of some metals and alloys in microgravity containerless experiments require such stringent gas purity levels. Copper based oxygen sorbents are not adequate for such specialized applications.

Cobalt is a transition series metal with some similarities to copper in oxidation and reduction behavior. From considerations of the particular position occupied by cobalt in the periodic table, it may be a good metal to try for an oxygen sorbent. Indeed, thermodynamic calculations indicate that the reduction of cobalt oxide by hydrogen is thermodynamically favorable and, in the reduced state, cobalt has a significantly better equilibrium oxygen removal than copper. The equilibrium oxygen removal values for copper and cobalt as a function of temperature are compared in Table 1. These values were obtained from the literature[4] values of free energies of species involved in the overall oxidation reaction and the relationship

$$\Delta G^\circ = -RT \ln K_p$$

where ΔG° is the free energy change for the reaction and K_p is the equilibrium constant.

While preparation of cobalt exchanged zeolites is relatively straightforward, reduction of such samples is more difficult. Samples of cobalt exchanged zeolites 13X and L were heated in a mixture of hydrogen and nitrogen at temperatures from 200 to 750 °C for periods up to 24 hours. In spite of such extensive treatment of the samples under a reducing gas, there was little indication of any significant reduction. In separate tests in a thermogravimetric analyzer (TGA), the cobalt exchanged zeolite samples were exposed to a reducing gas mixture consisting of 4% H2 - 96% Ar for 16 hours, purged in argon, and then allowed to contact oxygen at 600 °C. Monitoring of resulting weight change indicated that only a small amount of oxygen uptake occurred consistent with lack of reduction. Increasing the reduction temperature may speed up reduction process. However, a temperature of 600 °C is already beyond the stability limit for zeolite 13X and any further increase in temperature would be counter productive. For cobalt exchanged zeolite L, increasing the reduction temperature to 750 °C (which is close to its high temperature limit), did not help in reduction.

Literature indicates that reduction of certain metals is helped by platinum. For example, Sachtler[5], in his study of bimetallic catalysts, indicates that platinum may lower the reduction temperature of certain transition-series metals supported on a zeolite. Adding a small amount of platinum may then significantly alter the reduction characteristics of zeolites supported sorbents.

3 EXPERIMENTAL

Preparation of the catalytic Sorbent

The first step in preparation of the sorbents involves incorporating copper or cobalt into the zeolite by ion exchange. This is followed by addition of platinum to the cobalt exchanged zeolite by incipient impregnation.

Zeolites 13X and L were first treated with 1 molar solution of ammonium nitrate to replace sodium in the zeolite by ammonium. The ammonium exchanged zeolites were then treated with 0.1 molar solution of cobalt nitrate to replace the ammonium with cobalt. The cobalt exchanged zeolite, thus prepared, was well rinsed in distilled water and dried in an air oven at 250 °C. The addition of platinum to the treated zeolite obtained at the end of above step was carried out by addition of a platinum solution. The platinum solution was prepared by dissolving

Table 1. Equilibrium Oxygen Partial Pressures Resulting from
Formation of Copper and Cobalt Oxides

Temperature (K)	p_{O_2} based on CuO (atm)	p_{O_2} based on Cu_2O (atm)	p_{O_2} based on CoO (atm)
400	1.0×10^{-31}		
600	3.0×10^{-18}	1.8×10^{-22}	
700	2.0×10^{-14}	2.9×10^{-18}	
800	1.4×10^{-11}	4.1×10^{-15}	5.40×10^{-24}
900	2.1×10^{-9}	1.1×10^{-12}	1.37×10^{-20}
1000		9.4×10^{-11}	7.19×10^{-18}
1100			1.19×10^{-15}
1200			8.47×10^{-14}
1300			3.13×10^{-12}

Table 2. Composition of the Metallic Constituents in Various Zeolite
Supported Sorbents

Sorbent	Co (wt%)	Cu (wt%)	Pt (wt%)
COPT13X	5.2	---	0.1
COPTL	3.7	---	0.1
3COEX13X	5.5	---	---
COEXL	3.5	---	---
CU13X	---	6.0	---
CUPT13X	---	6.0	0.1

chloroplatinic acid, H_2PtCl_6, in distilled water. The addition of the solution was carried out in such a way so that all the solution was absorbed into the zeolite pores and no liquid solution was visible outside the pores. Further, the amount of platinum introduced this way was approximately 0.1 % by weight of the zeolite. Finally, the treated zeolite was dried again in the air oven at 250 °C. The copper exchanged zeolite sorbent was prepared as described in Reference 2.

The platinum-containing cobalt based sorbent preparations used in this study were COPT13X and COPTL which consisted of cobalt and platinum dispersed on zeolites 13X and L, respectively. For comparison, tests were also conducted with plain cobalt exchanged sorbents on these zeolites with no platinum addition, 3COEX13X and COEXL. In order to investigate the role of platinum in the reduction, limited tests were also conducted with copper exchanged zeolite 13X with platinum addition (CUPT13X) and without platinum (CU13X). The weight percentages of the metallic constituents on various zeolites are shown in Table 2.

Reduction of the Catalytic Sorbents

Reduction of the catalytic sorbents was carried out by heating the sorbent in a mixture of nitrogen and hydrogen at a temperature of 200 to 750 °C. The reactor system used for reduction is shown in Fig. 1. The hydrogen concentration in the reducing gas mixture was gradually increased from 1% H_2 to 20% H_2 over a 24 hour period. A reduction temperature of about 200-220 °C was used for copper-exchanged zeolite 13X, about 600 °C for cobalt containing zeolite 13X sorbents, and 600-750 °C for cobalt containing zeolite L sorbents.

Reduction of a few selected sorbents was also carried out in the sample holding pan of the Thermogravimetric analyzer. In this case, the cobalt exchanged zeolite beads were exposed to a reducing gas mixture consisting of 4% H2 - 96% Ar at the appropriate reduction temperature for 16 hours. After exposure to the reducing gas, the sample beads were purged in argon, brought to the desired temperature in argon environment and then contacted with oxygen. The uptake of oxygen was deduced from the sample weight increase with time.

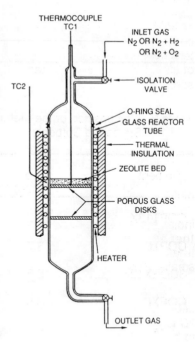

THERMOCOUPLE TC1

INLET GAS
N2 OR N2 + H2
OR N2 + O2

TC2

ISOLATION VALVE

O-RING SEAL
GLASS REACTOR TUBE

THERMAL INSULATION

ZEOLITE BED

POROUS GLASS DISKS

HEATER

OUTLET GAS

Figure 1 Quartz reactor assembly for reduction and oxidation tests

Transmission Electron Microscopy of the Reduced Sorbents

A single sphere of the reduced sorbent was used for the TEM analysis. A small wedge (less than 0.7 mm in the largest dimension) was cut from the edge as well as from close to the center of the sphere. These pieces were embedded in epoxy and cured overnight. Cross sections of these pieces were cut to 60-80 nm thickness using an ultramicrotome. These thin sections were supported on silver grids coated with a carbon substrate. The sections were then analyzed using a Philip's EM430 microscope at 200 kV. Bright field, dark field, and electron diffraction data were recorded.

Measurement of Oxygen Uptake

Oxygen uptake of the reduced sorbents was investigated by a measurement of the temperature rise upon exposing the reduced sorbent to a gas mixture of oxygen and nitrogen.

In order to measure the oxygen uptake of the cobalt-based sorbents, the sorbent bed was heated to about 625 ^oC and high purity nitrogen flowed through the sorbent at a rate of approximatelly 100 cc/min. When steady state conditions were achieved, oxygen at a rate of 10 cc/min was introduced into the nitrogen flow so that a gas mixture consisting of 10% O_2 - 90% N_2 flowed through the sorbent bed. The bed temperature recorded by the thermocouple was monitored as a function of time. For the copper-based sorbents, similar procedure was followed at sorbent bed temperatures in the range 90 - 200 ^oC.

4 RESULTS AND DISCUSSION

Reduction of the Cobalt Based Sorbent

When cobalt exchanged samples of zeolites 13X and L (with no platinum addition) were treated to a reducing gas at temperatures of 600-750 ^oC, no appreciable change in color of the zeolite beads occurred, indicating a general lack of reduction. This conclusion was supported by thermogravimetric analysis when sorbent beads, subsequent to treatment with a reducing gas, were contacted with oxygen and only a small oxygen uptake was observed.

However, when platinum-treated cobalt containing samples of zeolite 13X were subjected to the reducing gas at a temperature of 600 ^oC for 16 hrs in the reactor system of Fig. 1, the sorbent beads were found to be fully black in color. This indicated to a good measure that reduction had taken place.

Distribution of the Metal Crystallites in the Zeolite Framework

The distribution of the cobalt and copper crystallites in the zeolite framework subsequent to reduction was studied by Transmission Electron Microscopy.

For the cobalt based reduced sorbent COPT13X, Figure 2 shows the 'bright field' micrographs for a location near the edge of the sorbent bead obtained at magnifications of 70.9K and 128K, respectively. Similarly, Figure 3 shows the 'bright field' micrographs for a location near the center of the bead at magnifications of 41.7K and 70.9K, respectively. It is seen that the cobalt crystallites are distributed throughout the zeolite. A more detailed examination of these micrographs yielded the overall size range of these crystallites as well as the more narrow size range representing a majority (about 80 percent) of the crystallites. These results are summarized in Table 3.

For the copper based reduced sorbents CUPT13X and CU13X, TEM micrographs similar to those described above were obtained. Figure 4 shows the copper crystallites from the Platinum-free sorbent CU13X for a location near the edge of the bead at a magnification of 70.9K. Figure 5 shows the copper crystallites from the Platinum-added sorbent CUPT13X for a location near the center of the bead at a magnification of 34.0 K.

For CU13X, for locations near the edge, the presence of large clumps of crystallites which were aggregates of smaller crystallites was noticed. For CUPT13X, the analysis near the edge of the bead was made difficult due to the mobility of the copper crystallites. In fact, these crystallites were seen to move even as the analysis was being made. This is attributed to the partial oxidation of copper due to enhanced oxidation rate of copper by the presence of platinum. While some error must be allowed due to the mobility of the copper crystallites, the size ranges obtained for the crystallites are summarized in Table 3. The overall size range for the crystallites was 16 to 2600 Å. For reduced platinum-free sorbent CU13X, no similar mobility of the copper crystallites was observed. The crystallites were bigger with the overall size range from 190 to 3900 Å.

The TEM micrographs for the reduced copper based sorbents indicate that the addition of platinum results in smaller crystallites for copper. Smaller crystallites will lead to faster oxidation rate. Thus platinum addition will result in more efficient sorbents even in absence of direct participation of platinum in the oxidation. A similar effect of platinum for the cobalt based sorbents may be present. The crystallite size values in Table 3 do indeed show that the cobalt crystallites are generally below 150 Å in size. It was difficult to observe the cobalt crystallites in a sorbent without platinum addition because of the difficulty of reduction.

One possibility of a more direct participation by platinum in the reduction process is through chemisorption of hydrogen. The hydrogen

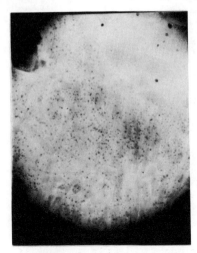

COPT13X - Edge
Magnification = 70.9K

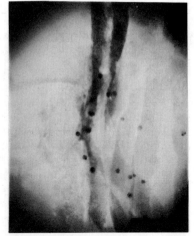

COPT13X - Edge
Magnification = 128K

Figure 2 TEM Bright Field micrographs for sorbent COPT13X
near the edge of the bead showing cobalt crystallites

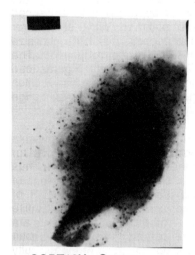

COPT13X - Center
Magnification = 41.7K

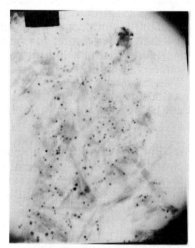

COPT13X - Center
Magnification = 70.9K

Figure 3 TEM Bright Field micrographs for sorbent COPT13X
near the center of the bead showing cobalt crystallites

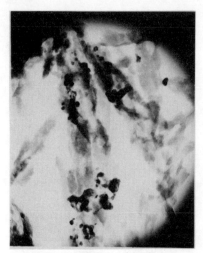

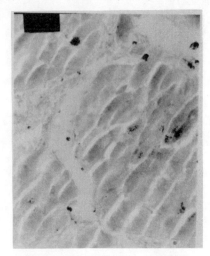

CU13X - Edge   X70.9K

CUPT13X - Center   X34.0K

Figure 4 TEM Bright Field
micrographs for sorbent CU13X
near the edge of the bead at a
magnification of 70.9 K

Figure 5 TEM Bright Field
micrographs for sorbent CUPT13X
near the center of the bead at a
magnification of 34.0 K

chemisorbed on platinum may be more accessible to copper in the
zeolite than hydrogen from ambient gas.

Oxygen Uptake by the Reduced Sorbent

The temperature-time profiles obtained for sorbent COPT13X and
the platinum-free cobalt-exchanged zeolite 13X (3COEX13X) are shown
in Figure 6. Similar profiles for sorbent COPTL and cobalt-exchanged
zeolite L (COEXL) are shown in Figure 7. The data in both Figures 6 and
7 clearly show a significant increase in bed temperature rise with the
platinum-containing material as compared to the plain cobalt exchanged
zeolite. A similar trend was obtained in oxygen uptake experiments with
copper-based sorbents CUPT13X and CU13X as shown in Figure 8.

The bed temperature rise upon contact with oxygen is a measure
of reactivity to oxygen. The area under the temperature rise-time plot is
proportional to the total oxygen uptake. As seen from Figures 6-8,
platinum-containing sorbents upon being subjected to the reducing gas
are a lot more reactive to oxygen than corresponding materials with no
platinum addition. The oxygen uptake of the platinum–containing
sorbents is significantly greater than the corresponding platinum-free
materials.

Above results lead to the conclusion that addition of a small
quantity of platinum to the cobalt or copper exchanged zeolites 13X and L
greatly enhances the reduction of the supported metal. From analysis of
the TEM micrographs, the presence of platinum diminishes the clustering

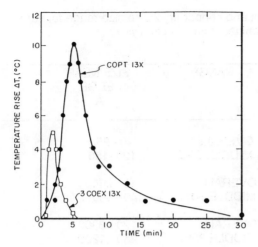

Figure 6 Bed temperature rise vs. time for cobalt supported on zeolite 13X with and without Platinum addition

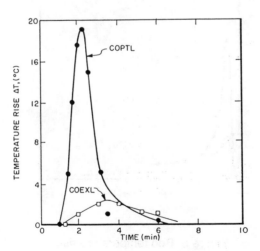

Figure 7 Bed temperature rise vs. time for cobalt supported on zeolite L with and without Platinum addition

Table 3. Cobalt and copper crystallite size range for various sorbents as determined from TEM analysis

SORBENT/ CRYSTALLITES	RANGE	SIZE NEAR EDGE Å	SIZE NEAR CENTER Å
COPT13X/ Cobalt	OVERALL MIDDLE 80%	40 - 250 120 -150	40 - 300 90 - 110
CUPT13X/ Copper	OVERALL MIDDLE 80%	16 - 2600 40 - 560	30 - 800 50 -100
CU13X/ Copper	OVERALL MIDDLE 80%	190 - 3900 250 - 1200	

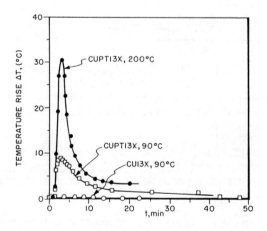

Figure 8 Bed temperature rise vs. time for copper supported on zeolite 13X with and without Platinum addition

of the crystallites and results in better dispersed and smaller crystallites. Such a configuration of the crystallites is consistent with higher reactivity. For the cobalt containing zeolites, the addition of platinum leads to relatively small, highly dispersed, and uniformly sized crystallites upon reduction. These are the desired attributes for high temperature oxygen sorbents of significantly increased efficiency and capacity.

5 CONCLUSIONS

The addition of a small amount of platinum by incipient impregnation to a cobalt or copper exchanged zeolite is seen to facilitate reduction and enhances the subsequent oxygen uptake rates and capacity. The improved performance appears to result from smaller and better dispersed metal crystallites formed during the reduction process. Platinum may play a more direct catalytic role in the observed oxygen uptake.

ACKNOWLEDGMENTS

The research described in this paper was carried out by the Jet Propulsion Laboratory, California Institute of Technology, under a contract with the National Aeronautics and Space Administration (NASA). This work is part of a comprehensive strategy for contamination control/management in NASA's Modular Containerless Processing Facility (MCPF) being developed for the Space Station Freedom. The author wishes to thank Mr. G.S. Hickey for the Thermogravimetric measurements.

REFERENCES

1. Sharma, P.K., and Seshan, P.K., "Trace Oxygen Removal by Copper Exchanged Zeolite Materials at Elevated Temperatures," in Proceed. Industry-University Advanced Materials Conference, edited by F.W. Smith (Advanced Materials Institute, Golden, Colorado 1989) pp. 856-863.

2. Sharma, P.K., and Seshan, P.K., "Activation of Copper Dispersed on a Zeolite for Oxygen Sorption," Chemically Modified Oxide Surfaces, Vol. 3, Editors Leyden, D.E., and Collins, W.T., Gordon and Breach, New York, 1990, pp. 65-80.

3. Sharma, P.K., and Hickey, G.S., "A Comparison of the Oxygen Uptake Characteristics of Copper Exchanged Zeolite with Copper Dispersed on a Silica support," Gas Separation & Purification, (1993) in Press.

4. Barrin, I., and Knacke, O., "Thermochemical Properties of Inorganic Substances," Springer-Verlag, New York, 1973.

5. Sachtler, W.M.H., Dossi, C., and Zhang, Z., "Zeolite Supported Bimetallic Catalysts", Paper No. 44c, 1989 Annual AIChE Meeting, November 5-10, San Francisco, California.

Catalysts for Environmental Control

Ronald M. Heck and Robert J. Farrauto
ENGELHARD CORPORATION, 101 WOOD AVENUE SOUTH, ISELIN,
NJ 08830-0770, USA

1   INTRODUCTION

The 90s are being called the "decade of the
environment". The public also has an increased
awareness of the necessity for controlling emissions
for preserving our planet. Governments around the
world are responding with increased regulations for
controlling emissions from mobile and stationary
sources. We are seeing an increased reliance on
catalysts to abate emissions from mobile and stationary
sources to meet increasingly more stringent
regulations.

Gasoline fueled automobiles have been equipped in
the U.S. with catalytic converters since 1976. Since
that time U.S. Federal emission standards required
reductions of CO to 3.4 g/mile, hydrocarbons (HC) to
0.41 g/mile and NOx to 1.0 g/mile (1981) for 50,000
miles. By the year 2004 further reductions to
1.7 g/mile for CO, 0.125 g/mile for HC and 0.2 g/mile
for NOx will be necessary.[1] To address these new
regulations it will be necessary for the gasoline
engine emission system to reduce emissions for the cold
start portion of the U.S. FTP test. In 1994 all
vehicles will have to meet regulations for 100,000
miles. In parallel automobile manufacturers want to
use less expensive Pd and/or base metal oxides to
replace the currently used more expensive combinations
of Pt and Rh.[2,3]

In 1994 select classes of diesel engines sold in
the U.S. will be using precious metal containing
catalysts to satisfy truck and bus regulations for
reductions of particulates required by the 1990
Amendment to the Clean Air Act of 1970.[4,5] Similar
requirements are expected in 1996 in Europe and 1997 in
Japan.

Compressed natural gas, and other energy sources
such as alcohols, are showing greater importance as

alternative fuels for powering vehicles. Precious
metal containing catalysts are being used to control
HC, CO and particulate emissions.[5]

In stationary applications such as power plants,
catalysts containing vanadium or zeolites are the
materials of choice for NOx while Pt containing
catalysts are used for abating CO emissions.[5] For
controlling chemical emissions from increasingly
"dirty" sources such as chlorinated hydrocarbons base
metal oxides[6,7] and precious metals[8] are being used.

Table I summarizes the major commercial examples
where catalysts are being used in controlling
emissions. The actual catalysts that are used in these
abatement processes are summarized in Table II. Note
that the catalytic materials range from precious metals
to base metals and combinations. This paper will
discuss the in use modification of these catalytic
surfaces in general, and then give actual commercial
experience for automotive, volatile organic compounds
and ozone.

2   CATALYST DEACTIVATION

During use of a catalyst for abatement of emissions
from a commercial installation, the catalyst is exposed
to an operating environment which modifies the surface.
A simplistic schematic of a catalytic surface is shown
in Figure 1. This represents a washcoated honeycomb.
The washcoat is deposited on the ceramic support
(honeycomb), and within the washcoat are the catalytic
elements, i.e. Pt, Pd, etc.. This structure is usually
dried and calcined before use in the final application.

However, during operation, the environment causes
changes in the surface due to exposure to higher
operating temperatures and impurities. This results in
changes in the catalytic surface as depicted in a
series of diagrams shown as Figure 2. More detailed
discussion on forms of catalyst deactivation can be
found in reference 9.

When a catalyst experiences excessive
temperatures, the following deactivation modes can
occur:

  • Sintering of the catalytic component and the
    carrier can result in the loss of active
    sites, causing a decline in overall
    conversion.

  • Rapid changes in temperature can result in
    thermal shock of honeycomb-supported
    catalysts due to differences in the

Table I: Commercial Examples of Catalyst for Environmental Control

- Three Way/Oxidation Catalyst
  - Gasoline Engines
  - CNG Engines
  - Flex Fuel Engines
  - Utility Engines
  - Off-Highway Vehicles

- Selective NOx Reduction
  - Lean Burn Stationary Engines
  - Power Plants
  - Process Heaters
  - Gas Turbine Co-Generation
  - Chemical Plants

- Non-Selective NOx Reduction
  - Natural Gas Re-Compression Engines
  - Engine Co-Generation Sets
  - Pumping Station Engines

- Oxidation Catalyst
  - Diesel Trucks
  - Diesel Cars

- Carbon Monoxide Abatement
  - Stationary Engines
  - Gas Turbine Co-Generation
  - Chemical Plants

- Volatile Organic Compound Abatement
  - Chemical Plants
  - Refineries
  - Manufacturing Plants
  - Ground Water Reclaimation
  - Food Processing

- Ozone Abatement
  - Aircraft
  - Sewage Treatment Offgas

Table II: Applications of Catalysts in Environmental Control

Application	Major Reactions	Active Catalytic Species*
Automotive (gasoline)	Oxidation of CO and HC's, and reduction of NOx	Mixtures of Pt, Pd and Rh with rare earth oxides
Stationary chemical facilities	a) Oxidation of CO and HC	Pt or Pt, Pd
	b) Oxidation of chlorinated HC's	Pt,V2O5 or unsupported Cr2O3 particulates
Lean burn stationary burners	a) Reduction of NOx with NH3	Pt; Pt, Au (low T), V2O5 (moderate T), Zeolites (high T)
	b) Oxidation of CO	Pt; Pt,Pd
High flying aircraft	Decomposition of O3 to O2	Pd; Co,Mn
Diesel truck and automobiles	Oxidation of soluble organic fraction, gaseous CO and HC	Pt, V2O5; Pd; base metal oxides
Stationary rich burn engines	NOx reduction and CO and HC oxidation	Pt, Rh

* Deposited on a high surface area carrier

Figure 1: Physical Representation Of Catalyzed Honeycomb Surface

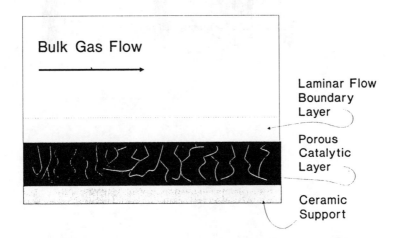

Bulk Gas Flow

Laminar Flow Boundary Layer

Porous Catalytic Layer

Ceramic Support

SINTERED CATALYST: AFTER AGING

SINTERING OF ALUMINA CARRIER

SINTERING OF PRECIOUS METAL

CERAMIC WALL OF MONOLITH

ATTRITION: AFTER AGING

STRONG CHEMICAL AND PHYSICAL BOND

CERAMIC OXIDE

SIMILAR THERMAL EXPANSION OF SUBSTRATE AND COATING

CERAMIC OXIDE MONOLITH

ATTRITION: AFTER AGING

- WEAK CHEMICAL AND PHYSICAL BOND DUE TO SMOOTH SURFACE
- CORROSION MAY OCCUR AT INTERFACE

DIFFERENT COEFFICIENTS OF EXPANSION

METAL SUBSTRATE MONOLITH

FRESH CATALYST: BEFORE AGING

EXHAUST GAS

ACTIVE PRECIOUS METAL SITE

HIGH SURFACE AREA CARRIER

CERAMIC WALL OF MONOLITH

POISONED CATALYST: AFTER AGING

INACTIVE PRECIOUS METAL COMPOUNDS

ACTIVE PRECIOUS METAL SITE

CERAMIC WALL OF MONOLITH

MASKING OF CATALYST: AFTER AGING

PLUGGING OF CATALYST PORES

COATING OF CATALYST SITE

CERAMIC WALL OF MONOLITH

Figure 2: Modes of Catalyst Deactivation

expansion/ contraction of the surface versus
the bulk of the substrate.

- Finally, differences in coefficients of
thermal expansion between the carrier and the
substrate can result in loss of adhesion of
the carrier to the support.

Two general classes of poisoning - selective and
nonselective - can result in catalyst deactivation.

Selective poisoning occurs when a feed compound
(i.e., not a reactant or product) specifically and
discriminately interacts with a specific catalytic
component, i.e. Pb reacts with Pt forming an inactive
alloy, resulting in a poisoning of the active species.
The process is selective in that the poison reacts with
the active catalytic component leading to a loss of its
activity. Since heterogeneous catalysts are usually
composed of a small amount of catalytic substance
deposited on a large-surface-area carrier, a small
amount of poison can rapidly deactivate the catalyst.
Thus, great care must be taken to understand thoroughly
the composition of all components in the feed and,
frequently, the chemical structure of the would-be
poisons.

Nonselective poisoning can be caused by a number
of phenomena, all of which are nondiscriminating in
that accumulation of foreign substances occur on both
the carrier and active catalytic opponents. This type
of poisoning frequently originates from impurities in
the gas stream condensing or depositing onto the
catalyst. An important form of nonselective poisoning
is often referred to as masking. Material entrained in
the gas phase, i.e. dust or aerosols, physically
deposits on the outer periphery of the catalyst and
blocks active site. With time this deposit may react
with active catalytic components, producing a less
active material. Also the impurity can deposit and
block access to the pores and thus prevents
accessibility to the active sites decreasing activity.

3   GASOLINE AUTO EMISSIONS CONTROL

Emissions from gasoline automotive exhausts are reduced
by catalytic converters located in the exhaust system.
Automobile exhaust converters were introduced in 1975
to oxidize the unburned CO and HC to carbon dioxide
(CO_2) and water (H_2O) in response to the original Clean
Air Act of 1970. A second generation converter was
introduced in the U.S. in 1981 to meet the 1 g/mile
nitrogen oxides standard as required in the Federal
Test Procedure and catalyzed the oxidation reactions
and simultaneously reduced nitrogen oxides to nitrogen.
These three-way converters (TWC) have been used in all

Figure 3: TWC Reactions

$$CH_n + (1 + n/4)O_2 \rightarrow CO_2 + n/2H_2O$$

$$CO + 1/2O_2 \rightarrow CO_2$$

$$NO + CO \rightarrow 1/2N_2 + CO_2$$

$$NO + H_2 \rightarrow 1/2N_2 + H_2O$$

$$(2 + n/2)NO + CH_n \rightarrow (1 + n/4)N_2 + CO_2 + n/2H_2O$$

$$CO + H_2O \rightarrow CO_2 + H_2$$

$$CH_n + H_2O \rightarrow CO + (1 + n/2)H_2$$

NOT ALL INCLUSIVE:
(N_2O FORMATION AT LOW TEMPERATURE)

(NH_3 FORMATION AT HIGH TEMPERATURE AND RICH)

Figure 4: Air-Fuel Ratio Determines Efficiency of Three Way Catalytic Converter

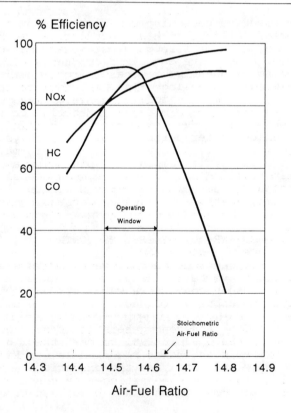

% Efficiency

Air-Fuel Ratio

new vehicles since 1981 in the United States, as well as Europe and Japan.

Catalytic materials in automobile converters are generally supported on a ceramic honeycomb monolith. The honeycomb, made of cordierite, contains 300 to 400 square channels per square inch, and is coated with an activated high surface area alumina layer called the washcoat. A five to one ratio of Pt to Rh (totalling between 20-40 g/ft^3) are highly dispersed on the washcoat. Three-way catalysts (TWC) operate near the stoichiometric air-fuel combustion ratio and at exhaust temperatures normally between 300 and 600°C, although higher temperatures are common. A second oxidation catalyst is sometimes used after the TWC for additional control of CO and HC emissions.

The reactions for the TWC catalyst are shown in Figure 3. These reactions are shown schematically in Figure 4 as a function of air to fuel ratio. Note that near the stoichiometric air to fuel ratio, the maximum conversion of all three components CO, NOx and HC occurs. The operating window determines the allowable change in air to fuel ratio around stoichiometric for a given performance. The actual deposition of the washcoat materials (stabilizers, base metals, etc.) and the precious metals automotive catalysis has become a good example of chemically modified surfaces. Figure 5 shows an example of a single coat automotive catalyst. Note the demarcation from the cordierite support to the single coat. Magnification is 700x for the SEM microphotograph. Figure 6 shows a double coat automotive catalyst. Here the SEM microphotograph is 300x and is in the corner location of a honeycomb channel. For this example, there is a bottom coat and top coat, indicating bulk segregation of the catalytic components and stabilizers.

Despite the twenty years of experience in automotive catalysis, catalyst longevity is still a key issue. This is particularly true if converters are to operate 100,000 miles. Reductions of lead in gasolines in the U.S. has minimized catalyst deactivation caused by lead. Other contaminants in the fuel and engine oil, especially phosphorous, zinc, and sulfur still present problems both as selective and non-selective poisons. Figure 7 shows microprobe analysis of a catalyst aged on an automobile for 50,000 miles. Analysis of the inlet section (Figure 7) of the catalyst indicates the presence of phosphorus, zinc and sulfur. Note that the phosphorus and zinc deposit near the surface of the washcoat, while sulfur deposits uniformly throughout the washcoat. The outlet section analysis (Figure 8) reveals a difference. Only sulfur is detected and at substantially reduced levels (10x's or smaller) when compared to the inlet section. These deposits modify the catalyst's activity. Phosphorus

Figure 5   Single coat automotive catalyst

SEM Microphotograph

Top Coat   Cordierite
                                                           Support

700×

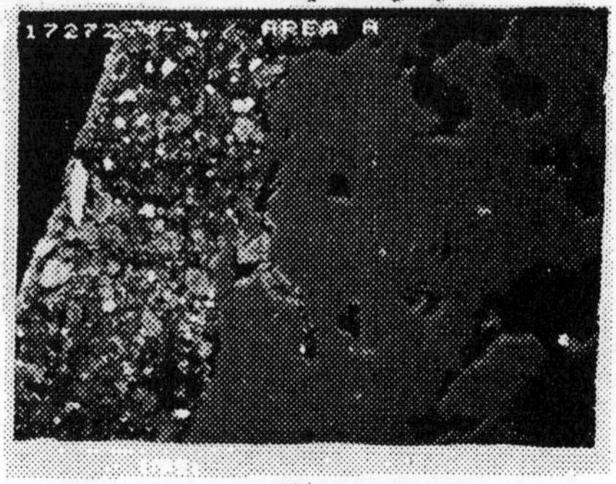

Figure 6   Double coat automotive catalyst

SEM Microphotograph

Top Coat

Bottom Coat

Cordierite
Support

300×

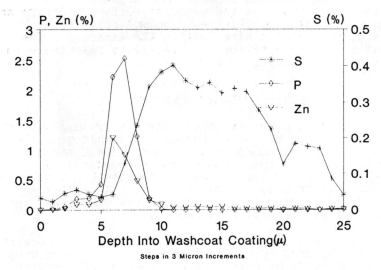

Figure 7: Deposits on Automobile
Catalyst-Inlet Section of Catalyst

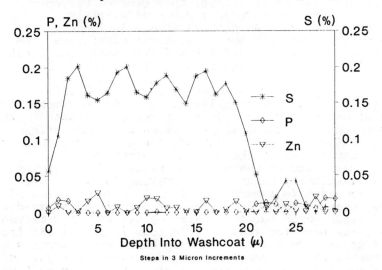

Figure 8: Deposits on Automotive
Catalyst-Outlet Section of Catalyst

and zinc compounds such as zinc dialkyl dithiophosphate in lubricating oil (a wear-retardant additive) reduce catalyst performance because they deposit (non-selectively) on the washcoat surface and form an amorphous glaze blocking exhaust molecules from reaching catalytic sites within the washcoat. The sulfur reduces activity by selectively reacting with the active catalytic sites.

4   VOLATILE ORGANIC COMPOUNDS

Manufacturing processes often involve organic solvents, feedstocks, or decomposition products that generate volatile organic compounds or VOCs. The Clean Air Act seeks to reduce emissions of these compounds by requiring businesses to install "reasonably available control technology" by May 31, 1995. An estimated 40,000 facilities, including printers, bakeries, and chemical plants will be affected.

Incineration, the most often used method, destroys these compounds by burning them at temperatures greater than 1000°C or by oxidation at temperatures between 300 and 350°C over a catalyst (catalytic incineration). Thermal (non-catalytic) incinerators are widely used even though their operating temperatures make their fuel costs relatively high. Their high-temperature operation also causes other problems: They need exotic high-temperature materials; they produce nitrogen oxides; and they sometimes yield undesirable by-products, such as dioxins from chlorinated materials. Their higher fuel costs often make them less desirable than catalytic incinerators for treating the fairly dilute gases typical of VOC emissions.

Catalytic incineration converts the volatile organics by catalytic oxidation at reduced temperatures to carbon dioxide and water. Since the nature of the emissions from chemical processing or incineration applications are often unpredictable, it is common to find impurities in the gas which often mask the catalyst. Due to the physical nature of masking, regeneration by selectively removing foreign contaminants is common practice. The formulation of the regenerating solution can by classified as acidic and basic, with varying chelating agents. A catalyst that was used for abating emissions in an off gas from the drying ovens used in aluminum and steel coating was returned after 3 years for regeneration. Analysis of the returned (as received) catalyst showed deposition of 0.3% P, 0.3% Sn, 0.007% Pb, and 2% Na contaminants, which originated from the cleaning products and pigments used in the process. The BET as-received surface area was measured to be only 3.55 m^2/g compared to 11 m^2/g for the fresh sample. When subjected to a standard activity test, this catalyst gave only 13%

toluene conversion.   Toluene here is used as a model
compound to simulate performance in the installation.

Two different chemical regeneration treatments
were evaluated. Figure 9 compares the toluene
conversion of the cleaned catalyst with that for a
fresh catalyst. After acid treatment, the activity was
improved over that for the as-received catalyst, but
was still significantly lower than the fresh catalyst
activity. For this cleaned catalyst, conversion in the
bulk mass-transfer regime (>300°C) was only 63% versus
86% for a fresh catalyst under similar laboratory test
conditions. Apparently, a contaminant layer
(non-selective poisoning) had remained on the surface
to block accessibility to a portion of active sites.
Wet chemical analysis also showed that most of the
deposits had not been removed by this chemical
treatment. The alkaline-treated catalyst was found to
be much more active. As shown in Figure 9, about 90%
of the original catalyst activity could be recovered by
this cleaning method. Wet chemical analysis indicated
that most contaminants were removed from the catalyst.
The contaminant levels after cleaning were 0.03% P,
0.03% Sn, 0.001% Pb and 0.4 Na. The BET surface area
after this treatment was increased to 9.6 m^2/g from
3.5 m^2/g. This confirms that higher temperatures were
not the cause for the catalyst deactivation; rather,
deactivation was caused by masking of the outer surface
and pore plugging.

Another example is abatement of the emissions in
the effluent from a water-base paint area. During
operation, the reactor temperature could not be
properly controlled and a decline of catalyst
performance was observed. The catalyst was found to
contain large deposits of unburned hydrocarbons and
phosphorus (1.12%). The laboratory test results shown
in Figure 10 indicate a low toluene conversion activity
for this catalyst (as received). After treatment with
an alkaline solution, the cleaned catalyst gave the
same toluene conversion response as a fresh catalyst.
The phosphorous content was reduced to 0.024%. At
operating temperatures in excess of 300°C deposition of
the hydrocarbons was prevented and the catalyst
activity did not decline. In addition, periodic
catalyst regeneration by chemical cleaning enables the
system to maintain its activity by removing the
deposits of phosphorous compounds.

In another application, Pt/Al_2O_3/honeycomb
catalyst system (55 cpsi) was used to oxidize solvent
vapors including butyl acetate, xylene and butane, and
hexamethyl disiloxane (HMDS) aerosols from a negative
photo electric process in a semiconductor line. The
catalyst activity dropped rapidly. The activity for
toluene oxidation of this deactivated catalyst was
tested before and after chemical cleaning. Figure 11

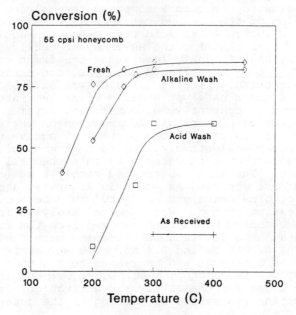

Figure 9: Toluene Conversion-Aged VOC Catalyst Before and After Regeneration: Metal Strip Coating

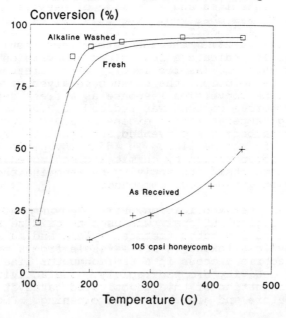

Figure 10: Toluene Conversion-Aged VOC catalyst before and after Regeneration: Automobile Coating

Figure 11: Toluene Conversion-Aged VOC Catalyst Before and After Regeneration: Electronic Component

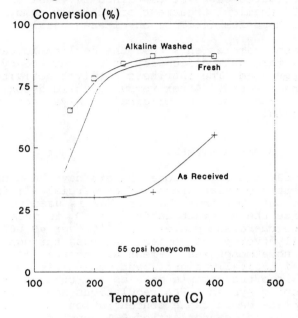

Figure 12: Ozone Abater Performance
(20,000 Flight Hours)

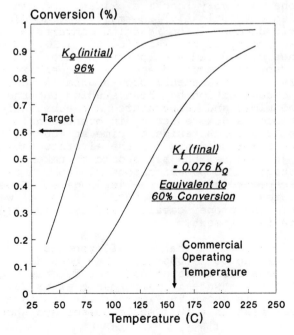

shows that the as-received catalyst exhibited a very
low toluene activity. However, the alkaline-cleaned
catalyst gave the same performance as a fresh catalyst.
The deactivated catalyst contained 27.4% Si on the
surface.   Chemical treatment effectively removed most
of these deposits (4.2% Si on cleaned catalyst).

These examples show that the decline in catalyst
activity through modification of the surface by masking
can be reversed. The intrinsic catalyst activity had
not been affected. After removing these deposits by a
chemical treatment, the original catalyst activity can
be restored.

    5   OZONE ABATEMENT

Since 1983, catalytic ozone abaters have been used by
the aircraft manufacturers and commercial airlines to
remove ozone (O_3) from the air that is used to
pressurize the aircraft cabin. The O_3 is present in
the troposphere and planes that fly over 40,000 feet,
especially over polar routes, can pass through these O_3
containing atmospheres. Fresh air containing small,
but physiologically significant, quantities of O_3 is
brought into the cabin's make-up air through the air
conditioning system. FAA regulations now require that
the airplane cabin ozone concentration (on a time
weighted basis) cannot exceed 0.1 vppm (sea level
equivalent).[10]

Ozone is a very reactive oxidant and can be
decomposed catalytically, thermally, or adsorbed.   The
exothermic decomposition reaction converts ozone to
oxygen according to the reaction $2O_3 \rightarrow 3O_2$.   Early
approaches considered adsorption, however, the weight
of the adsorbent required and the frequent regeneration
necessary (or replacement) proved unsatisfactory.
Thermal decomposition has been studied but the
residence times and temperatures are substantial.[11]
For instance, a 90% reduction in ozone concentration
requires 10 seconds residence time at 600°F (316°C).
Since the compressed air for the aircraft cabin comes
off the jet engine, the air should be taken off the
lowest stage of compression possible to reduce
parasitic power loss from the jet engine. Thus,
temperatures closer to 300-400°F (150-200°C) were more
suitable. The proper catalyst and system design can
meet this requirement.

Performance and analysis of ozone abaters with
flight hours up to 25,000 hours has been determined to
identify possible modes of catalyst deactivation. An
example of the change in performance after 20,000
flight hours is shown in Figure 12. This is a
laboratory test on a full scale abater and shows the

change in performance as a function of operating temperature.

Analyses of abaters having 10,000 and 25,000 flight hours were completed. Surface deposits of sulfur, phosphorus and silicon were noted as shown in Figures 13, 14 and 15. Furthermore, at 10,000 flight hours there is an axial concentration profile for all three components. This concentration profile is much less pronounced for sulfur and phosphorus at 25,000 flight hours. The concentration levels are much higher for phosphorus when comparing the different operating times with only minor differences in the levels of silicon and sulfur.

Porosity measurements were also determined using nitrogen adsorption/desorption curves. The porosity data of the bulk catalyst (Table III) indicated that the catalyst with 25,000 flight hours contained about one half of the porosity of the catalyst having 10,000 flight hours. The mean pore radius was also noted to be reduced as the catalyst aged.

Based on the pore size analysis of the catalyst, it is apparent that masking is the primary cause of deactivation. The masking causes pore plugging reducing accessibility of the ozone to the palladium sites within the washcoat. This pore plugging is caused by the deposit of phosphorus and silicon containing compounds. The major source of silicon and phosphorus is from the lubricating and hydraulic oils widely used in the aircraft.

It is believed that the sulfur is a non-selective deposit and is chemically bound with the washcoat material.

6   CONCLUSION

Catalysis is certainly an example of a chemically modified surface which promotes specific reactions. In use, this surface is further modified resulting in a change in the performance of the catalyst. By understanding how this surface is modified in an application, new catalysts can be designed or regeneration methods devised to restore the intrinsic surface activity.

7   ACKNOWLEDGEMEMT

The authors wish to thank Dr. Harold N. Rabinowitz for providing background information and James Rogalo for his artwork.

Figure 13: Phosphorus Analysis

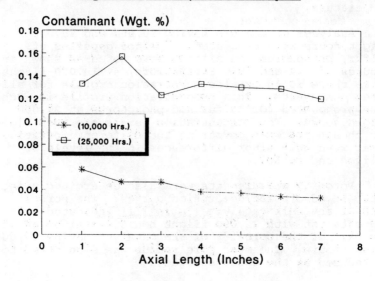

Figure 14: Sulfur Analysis

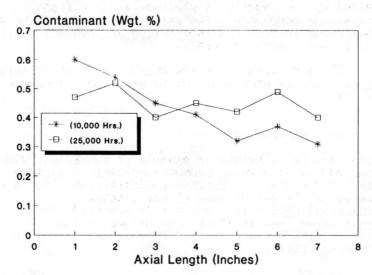

Figure 15: Silicon Analysis

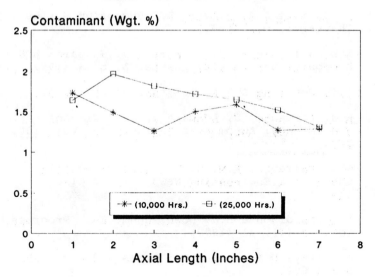

Table III: Comparison of Porosity
Flight Aged Catalyst

Flight Hours	0	10,000	25,000
BET Desorption			
Porosity (cc/g)	0.2469	0.0852	0.0318
BET Area (m^2/g)	98.8	32.6	18.7
Mean Pore Radius (Å)	49.4	53.2	37.9

REFERENCES

1.   J. Summers & R. Silver; ACS Symposium Series 495,
      Chapter 1, p.2 (1992).

2.   W.B. Williamson, J. Summers & J. Scaparo; ACS
      Symposium Series 495, Chapter 3, p.26 (1992).

3.   J.C. Dettling & Y.K. Lui; SAE 920094, 2/92.

4.   R.J. Farrauto, J. Adomaitis, J.A. Tiethof &
      J.J. Mooney; Automotive Engineering, Vol. 100,
      2/92.

5.   R.J. Farrauto, R.M. Heck & B.K. Speronello;
      Chemical & Engineering News, Vol.70, (36), 34
      (1992).

6.   L.C. Hardison & E.J. Dowd; Chem. Eng. Progress,
      Vol. 73 (7), 31 (1977).

7.   J. Berty; U.S. Patent 5,021,383 and U.S. Patent
      5,114,692.

8.   G. Lester; International Patent 90/13352, also
      presented at the 82nd Annual Meeting of the Air
      and Waste Management Association, 6/89.

9.   J. Chen, R.M. Heck & R.J. Farrauto; Catalysis
      Today, Vol. 11, 517 (1992).

10.   R.M. Heck, R.J. Farrauto & H.C. Lee; Catalysis
      Today, Vol. 13, 43 (1992).

11.   A.E. Axworthy & S.W. Benson; ACS, 3/59.

Mechanism of Surfactant-assisted Increase in Coal Liquefaction Yields

Gregory S. Hickey and Pramod K. Sharma
JET PROPULSION LABORATORY, CALIFORNIA INSTITUTE OF
TECHNOLOGY, PASADENA, CA 91109, USA

1 ABSTRACT

The addition of sodium lignosulfonate surfactant to coal during liquefaction batch autoclave tests increases the coal liquefaction process yields. Analytical tests conducted on selected process runs with and without surfactant show that the surfactant acts by a mechanism of preventing the agglomeration of coal particles and facilitating the breakage of crosslinks of the coal into smaller fragments.

2 INTRODUCTION

The liquefaction of coals is a promising technology for producing alternate fuels that may eventually replace petroleum based fuels. It has long been known that the operating conditions (such as solvent type and structure, the hydrogen to carbon (H/C) ratio, temperature, etc.) play a significant role in the dissolution of the organic matter in the coal. The possible effects of lowering the viscosity and surface tension of the liquid phase in the reactor have mostly been speculated upon but not systematically investigated. This present work studies the effect of adding a surfactant to the coal liquefaction process and the mechanism of its action.

In the simplest terms, the coal liquefaction process is the hydrogenation of coal in a coal derived liquid solvent with high pressure hydrogen at elevated temperatures. The present state of the art is the use of two stage ebullated bed reactors, with supported catalyst in one or both of the reactors. The reactor operating conditions are typically 425°C with 2000 psig hydrogen, and residence times of one to two hours. A fraction of the coal derived liquid is used as recycle to sustain the process. Process yields approach 90 to 95% depending on the coal, with direct liquid conversions (C$_4$ to 525°C distillates) of 75% achievable. However, the process can be conducted completely thermally and

produce   carbon conversion yields over 80%.   A review
of the chemistry of   the thermal processes has been
published by Whitehurst, *et.al.* (1).

The structure of coal has been investigated by many
researchers and it is generally agreed to consider coal
as   a   highly   crosslinked   amorphous   polymer,   which
consists   of   a   large   number   of   stable   aggregates
connected by relatively weak crosslinks. Figure 1 shows
one possible structure.   At high temperatures in the
presence of a hydrogen donor solvent, the fragments will
break up into radicals that are capped by the solvent to
form stable species before reaction.   These species
consist of aggregates, or groups of aggregates with
molecular weights   ranging from 300 to 3000.   In the
absence of a hydrogen donor solvent, the radicals will
react to form either char or coke.   These coal derived
fragments are primarily preasphaltene and asphaltene
particles.   They consist of high molecular weight
polyaromatic or polycyclic molecules with links to
heteroatoms of N, O and S.

The coal liquefaction process occurs in three
stages: 1) the solubilization and breakage of the
crosslinks of the coal in the solvent into smaller
fragments, 2) the reaction of hydrogen   with the coal
fragments by hydrogen transfer to lower molecular weight
molecules, and 3) the rehydrogenation of the solvent.
All three processes are important to occur to have a
functional,   self-sustaining   continuous   industrial
process. The three primary considerations for the
sustainablilty and high reactivity of the system are the
hydrogen donor capability of the   of the liquid solvent
produced,   the   physical   solubilization   of   the   coal
particles within the solvent, and the hydrogen transfer
for reaction within the system.

3 SELECTION OF THE SURFACTANT AND ITS EFFECT

Coal fragments typically disperse poorly in
nonpolar and mildly polar solvents, and tend to
agglomerate into aggregates of high molecular weight.
The motive for adding a surfactant with an "asphaltene-
like" structure was to better disperse the particles and
prevent them from aggregating.   Sodium lignosulfonate
surfactant was chosen because it is an oil-compatible
colloidal surfactant that is commercially available as
an inexpensive by-product from   waste   paper and pulp
processing.   It has an approximate structure shown in
Figure 2. It has a molecular weight in the 300 to 1000
range. At mild processing temperatures it   readily
disperses in hydrocarbon solvents as molecular units. It
is typically used in industry as a dispersion agent for
solids, and as an oil-water emulsion stabilizer.   It is
thermally stable at the processing temperatures and
pressures used in coal liquefaction. The lignosulfonate
may act as an aid in dissolution of the coal particles

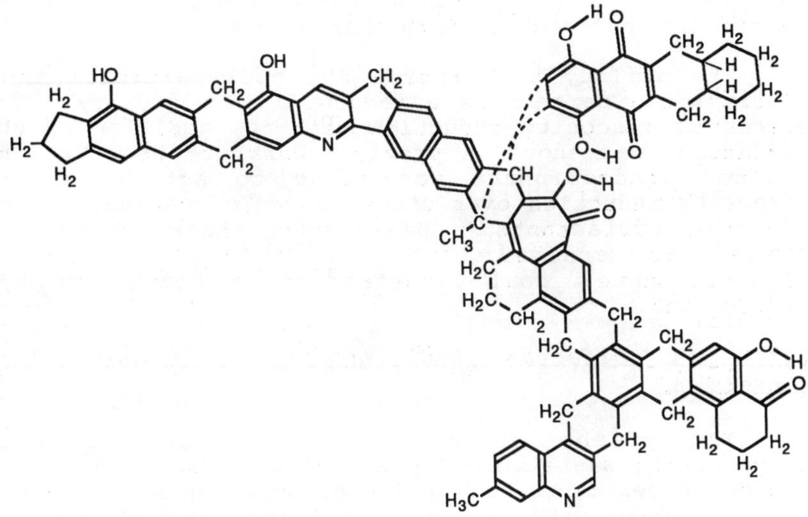

Figure 1: Representative structure of coal.

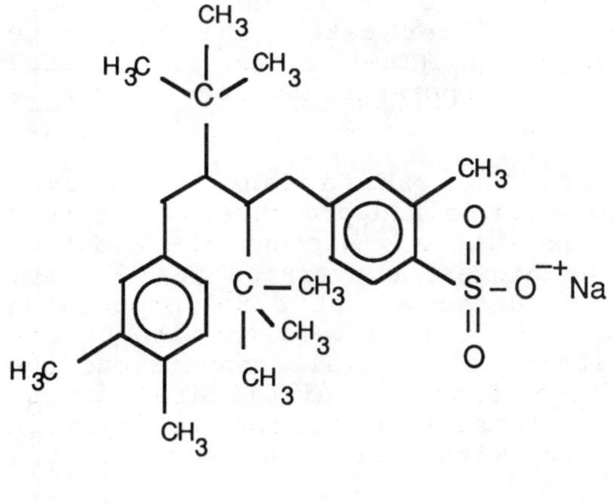

Figure 2: Representative structure of sodium lignosulfonate surfactant.

into asphaltene and preasphaltene units, as micelles and colloids, without the formation of radicals which can recondense and react to form char or coke.

The ability to disperse the coal particles in a slurry would be seen as a reduction in viscosity. The effect of viscosity reduction with the addition of the surfactant was shown in previous work by Hsu (2). The sodium lignosulfonate was found to act better in viscosity reduction over other anionic, cationic, and nonionic surfactants. Based upon these results, a systematic analysis of the effect of sodium lignosulfonate coal liquefaction on conversion was carried out.

Effect Of Surfactant Addition On Coal Liquefaction Conversion

The effect of surfactant addition on coal liquefaction was studied in a series of batch autoclave reactor experiments. Experiments were conducted in a 1 liter reactor with 100 g of Illinois # 6 coal and 200 ml of SRC-2 recycle solvent obtained from Hydrocarbon Research Incorporated (HRI). The sodium lignosulfonate surfactant was added in concentrations of 0.5, 1.0 and 2.0 wt% additions. The processing temperatures were varied from 300 to 375°C, with hydrogen pressure at 1800 psig, and a processing time of 1 hour. Conversions on a moisture and ash free (MAF) basis were determined by a mass balance of the amount of coal derived liquid produced and coal-slurry filter cake that could be extracted by tetrahydrofuran (THF) after processing. Details of the experimental apparatus and results can be found in Reference 3.

The addition of the surfactant increased conversions at all temperatures and pressures. Figure 3 shows the MAF conversions with and without the addition of the surfactant. The highest conversion was 92% for the addition of 2 wt % surfactant at 375°C processing temperature, compared to 83.1 % conversion without the surfactant. In addition to the overall conversion, the amount of light boiling fraction distillate (up to 300°C) was determined. Figure 4 shows the increase in light boiling fraction with the addition of the surfactant.

4 INVESTIGATION OF THE SURFACTANT MECHANISM

Figure 5 shows one possible mechanism for the reaction of the coal with the surfactant. Without the surfactant, the coal would agglomerate and only the outer peripheral groups would react. This would lead to increased gas production and lower conversion. If the preasphaltene coal fragments are separated by the

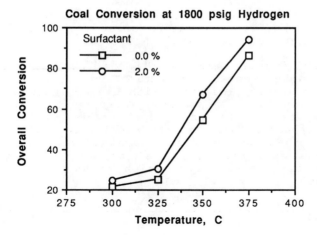

Figure 3: Temperature dependence of coal conversion,
with and without surfactant.

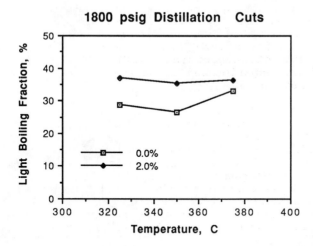

Figure 4: Temperature dependence of light boiling
fractions, with and without surfactant

H
R
O
O
H
R'
+
O
-
O
S
R"
O
R"
O
S
O
O
-
+
H
R
O
+
O
H
R'

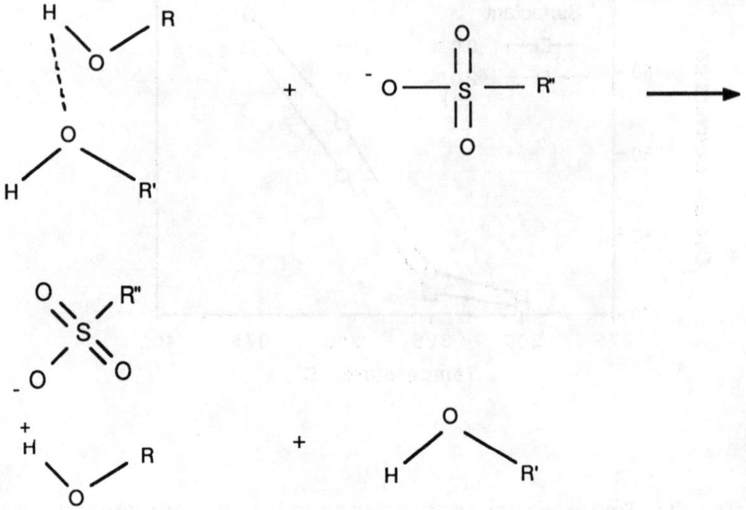

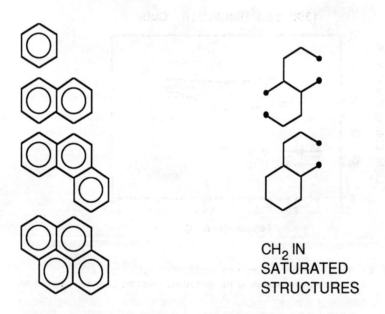

Figure 5: Representative mechanism for the role of the surfactant.

CH2 IN
SATURATED
STRUCTURES

Figure 6: Representative functional groups of coal liquids that were seen by NMR.

surfactant, larger coal fragments could react with the hydrogen and give higher conversion.

To further evaluate the role of the surfactant, a series of analytical evaluations were made of the processed coal filtercake, filtrate and residue. The primary objectives of the analytical work were to determine the distribution of the surfactant in the product streams, and to quantify changes in the liquefaction product due to the addition of the surfactant.

Sodium Analysis on Filtercake and Extracted Residue

Two residues from 375°C processing runs, one with 2% surfactant, and the other without any surfactant were analyzed by SEM/EDS to determine if any surfactant was retained in the residue after successive extractions with hexane, toluene and THF. The surfactant contains 7.0 % sulfur by weight and 5.9% sodium by weight. Since the coal has 4.5% sulfur initially, the presence of sodium was used to qualify the presence of surfactant. Sodium was found evenly dispersed in the residue of the surfactant added sample by SEM/EDS analysis. Flame atomic absorption was conducted on the two residues to determine the total sodium. From a mass balance based on the increase in sodium content of the samples where the surfactant was added, it was determined that 53% and 47% of the surfactant was retained in the filtercake and extracted residue, respectively.

FTIR Analysis

The filtrates, filtercakes and residues were characterized by Fourier Transform Infrared Spectroscopy (FTIR) to determine structural differences between samples with and without surfactant. The 0% and 2.0% surfactant added runs at 375°C were analyzed by the diffuse reflectance method, as were the pure surfactant and recycle solvent. The surfactant was seen in the 2.0% sample of the filtrate, filtercake and residue as a distinctive sulfonate group stretch at wavenumber 1145 cm^{-1}. The filtrates, filtercakes and residues were all multifunctional group structures. With the exception of the surfactant presence, there was only a little difference between the two filtrates and residues. The filtercake from the surfactant added sample showed some additional C-H stretching at around 2900 to 3000 cm^{-1} and some additional aromatic stretches near 1000 cm^{-1}. Since there was a significant difference in the light boiling fractions of the filtrates from the 350 °C runs, the 0% and 2.0% surfactant added filtrate products were analyzed by FTIR. These two samples showed more differences than the 375°C filtrates. The filtrate corresponding to 2.0 % surfactant showed much stronger C-H stretches and aromatic stretches than the 0%

surfactant added filtrate product. FTIR analysis was also conducted on successive hexane, toluene and THF extracts from the 0% and 2% surfactant added filtercake products. No significant structural differences were found between the extracts with and without the surfactant.

NMR Analysis

Proton NMR and ^{13}C NMR were conducted on the two 350°C filtrates. There two structures were very similar, consisting of aliphatic, polycyclic and polyaromatic compounds. While no clear structural differences with and without the surfactant were observed, the analysis indicated that the majority of the aromatics were of the structure shown in Figure 6.

GC/MS Analysis

To better quantify the filtercake products, gas chromatography (GC) was conducted on successive hexane, toluene and THF extracts from the 0% and 2.0% surfactant added filtercake products from the 375°C processing temperature runs. Samples of the extracts were injected into the GC (a 25 m length DB5 capillary column) and separated by the column. Selected compounds were routed to a mass spectrometer for species identification. Reference samples of C_{10} to C_{30} compounds were also tested to provide a baseline for determining average molecular weights.

The hexane extracts were mainly saturated and unsaturated hydrocarbons. The extract from the 2% surfactant added product showed the presence of a larger number of species and an average molecular weight of 250 compared to a molecular weight of 230 for the extract without surfactant. The toluene extracts were primarily saturated and unsaturated hydrocarbons with some aromatic species. Both samples had an approximate molecular weight of 260. The THF extracts were mostly aromatic species. Though these species had a longer column retention time, they had a lower average molecular weight. The surfactant added THF extract had an average molecular weight of 150 compared to 125 for the 0% extract. There was good qualitative agreement regarding species structure between the GC/MS and FTIR analyses.

XPS Surface Analysis of the Filtercakes

Surface analysis of the hexane-washed filtercakes from the 0% and 2% surfactant added processed coals was conducted using X-ray Photoelectron Spectroscopy (XPS). The technique analyzes only the top 50 Å of the surface. The analysis indicated only a trace level of the sodium on the surface for both the samples. This result is at odds with the quantitative sodium determination by atomic absorption. However, a possible explanation is that a majority of the surfactant migrated to macropores beneath the top surface. As macropore surface area is expected to be significantly greater than the external geometric surface, it is conceivable that the sodium concentration on the external surface dropped below the detection level of the XPS.

Surface Morphology by SEM Analysis

The surface morphology of the coal after processing was examined by SEM. Samples of the filtercake were extracted with hexane to remove absorbed oils. Examples of the surface morphology are shown in Figure 7. The processed coal where surfactant had been added has a more open structure, with greater surface area than the coal processed without surfactant. The addition of the surfactant appears to have opened up the pore structure of the coal, producing larger macropores. This provides evidence that the surfactant action is being facilitated by opening up the structure of the coal during processing. The coal hydrogenation occurs at the surfaces (internal pore and external) of a coal particle. If one considers the reactions on a macropore surface, in the presence of the surfactant, such reactions will be significantly enhanced due to the open structure of the coal fragments. Thus the macropore will be enlarged relative to its size (or opening) when no surfactant is present. This enlarging of the macropore was observed from the SEM morphological analysis of the hexane washed filter cakes.

5 DISCUSSION

The batch autoclave tests indicate a significant increase in overall coal conversions at all operating temperatures in the range from 300 to 375°C. The increase in light boiling fractions of the filtrate, on the other hand, is significant up to 350°C only. Analytical test results on the structure and functional groups of organic compounds in the filtrate, filtercake and extracted residues show only minor differences in the samples processed with and without surfactant. The same is also true of the structure and functional groups for the successive extracts obtained by washing the filtercakes with hexane, toluene, and THF. The only significant differences obtained for the hexane and THF extracts are in the average molecular weights, which were higher in the 375°C processing run with 2.0 % surfactant than for the case with no surfactant added.

The above results suggest that this surfactant does not significantly alter the three stage progression of coal liquefaction described earlier. The surfactant appears to speed up the first stage, the breakage of the

7a)

7b)

Figure 7: SEM micrographs of hexane extracted
          filtercakes a) with and b) without the
          surfactant added during processing.

crosslinks in the coal. If this process was rate controlling, the remaining two stages, i.e., the reaction of the hydrogen with coal fragments and the rehydrogenation of the solvent, would also speed up.

The following discussion will help visualize how the surfactant may help in the breakage of the crosslinks of the coal. Let the coal be represented by R-O-H. Without the surfactant, coal molecules (represented as R-O-H and R'-O-H in Figure 5) are associated by hydrogen bonding between the H-atom of one coal molecule and the O-atom of the other. Due to its anionic polar nature, the surfactant attaches itself to the acid site (H atom) on the coal. Thus it breaks the association with the O-atom between different coal molecules, increasing the favorable hydrogenation reactions. Because the lignosulfonate is a Lewis acid and a partially hydrogenated aromatic compound, it is possible that it could act as a hydrogen donor in the reaction. This is a possible additional avenue of increase in favorable liquefaction processes due to the surfactant.

The addition of the surfactant at 375°C processing temperature increased coal conversion from 83% to higher than 90%. The increase was primarily related to the increase in the THF extracted species from the filtercakes. As the THF extracted species are expected to be preasphaltenes and the GC/MS analyses indicate that these species are polyaromatic compounds, the increase in the average molecular weight of the THF extract was not surprising. The species that converted to a THF soluble product due to the surfactant are expected to be bigger than those that did not require the surfactant.

The effect of the surfactant on the quality of the filtrate (i.e. the light boiling fractions) are clearly dependent on processing temperature. The surfactant increases the straight chain (lower boiling) hydrocarbons (C-H stretch) as well the molecular weight of the polyaromatics. At 350 °C, the increase in straight chain hydrocarbons appears to dominate, resulting in an increase of lighter fractions. At 375°C, the production of aromatics appears to counter the production of aliphatic hydrocarbons. Hence, a significantly smaller increase in lighter fractions is observed.

6 CONCLUSIONS

The observed increase in the overall coal conversion due to the addition of the surfactant, sodium lignosulfonate, appears to result mainly from the breakage of crosslinks of the associated coal molecules. This, in turn, appears to lead to greater access of hydrogen to the coal fragments and thus a significant increase in the hydrogenation rate and subsequent solubilization rate.

7 ACKNOWLEDGMENTS

The research described in this paper was carried out by the Jet Propulsion Laboratory, California Institute of Technology, under a contract with the National Aeronautics and Space Administration. The work was sponsored by the Pittsburgh Energy Technology Center, Department of Energy, through DOE/NASA Interagency Agreement No. DE-AI22-92PC92150. The authors gratefully acknowledge Mark Anderson and Gary Plett for their assistance in the analytical work.

8 REFERENCES

1. D. Whitehurst, T. Mitchell, and M. Farcasiu, 'Coal Liquefaction, The Chemistry and Technology of Thermal Processes', Academic Press, New York, (1980).

2. G. Hsu, 'Surfactant Studies for Coal Liquefaction, Final Internal Report', Submitted to the U.S. Department of Energy, Pittsburgh Energy Technology Center, Contract No. DE-AI22-89PC882, (December 1990).

3. G. Hickey, P.K. Sharma, 'Surfactant Studies for Bench-Scale Operation, Second Quarterly Technical Progress Report', Jet Propulsion Laboratory Publication 93-7, (March 1993).

The Effect of Glass Fiber Surface Coatings on Fiber Strengths and the Distribution of Flaws

J. A. Gómez and J. A. Kilgour

OSI SPECIALTIES, TARRYTOWN TECH. CENTER, TARRYTOWN, NY 10591, USA

1 INTRODUCTION

The effective design and development of new silanes for sizing or finishing glass fibers requires an understanding of the mechanisms that contribute to the desirable mechanical properties of glass reinforced composites. A variety of studies[1] have shown that organofunctional alkoxy silanes under commercial sizing conditions are capable of reacting at the glass surface to create the glass/sizing interface. Further, there is usually a sufficient excess of silane present to form an extensive interphase region that interacts with both the initial interfacial layer and the matrix resin of the composite. The interface and the interphase layers play dominant roles in determining the initial glass fiber tensile strength and the interfacial shear stress transmission critical to the ultimate mechanical properties of the reinforced composite. The synthetic challenge is to manipulate chemical structures in order to both retain the tensile strength of the glass fibers and to optimize the transfer of stress on the composite from the matrix to the glass.

Organofunctional alkoxysilanes are designed such that the alkoxysilyl portion will hydrolyze in solution and then condense during curing with silanols on the glass surface and with each other to form a covalently bonded interfacial layer and interphase network. Altering the nature and number of alkoxy groups will change the rate, extent and characteristics of the siloxy networks that are produced. An organofunctional group is generally attached at the other end of the molecule to interact with the resin matrix to form a covalently bonded silane/resin interphase. To this end, a wide range of reactive organofunctional groups have been synthesized including aminoalkyl, ureidoalkyl, vinyl, methacryloxypropyl, and epoxyalkyl to name a few. Among these, aminoalkyl functional silanes were selected for this study due to their commercial importance, and because of their unique property of catalyzing the attachment and condensation reactions as well as their ability to react with epoxy resin systems.

Historically, several test methods have been developed to measure the physical performance properties of both glass fibers and glass reinforced composites. Glass fiber tensile strengths are measured directly at different gauge lengths , then statistically interpreted and mathematically described as bimodal distribution termed a "double box" distribution[2]. Measuring the more complex performance parameters relating to interfacial shear stress transmission in reinforced composites has been performed by several methods including the pull out test[3], the push out test[4] and the microbond test[5]. Although all have their limitations, perhaps the most useful is the embedded single fiber test[6], which has been adopted for this study and employed over a range of temperatures. Determining the change in interfacial shear stress transmission with temperature has provided insight into the mechanisms of silane performance and has generated useful information for commercial applications where withstanding temperature changes is critical.

2 EXPERIMENTAL

The organofunctional silanes used in this study are the commercially available OSi Specialties Organofunctional Silanes A-1100, 3-aminopropyltriethoxysilane, and A-1170, bis-(3-trimethoxysilylpropyl)amine, and the experimental silanes, 11-AUTMS, 11-aminoundecyltrimethoxysilane, 3-APMDS, 3-aminopropylmethyldimethoxysilane, and 4-ADBMDS, (4-amino-3,3-dimethyl)butylmethyldimethoxysilane. The silanes were applied to the glass fibers as 2.5 wt% solutions in water acidified with acetic acid to pH 4.0. No film-formers or other ingredients were included in the size.

The silane size solutions were applied to freshly drawn glass fibers. E-glass marbles, obtained from Schuller International, were melted at 1200°C and drawn from a 204 hole glass fiber bushing. A single fiber was isolated and pulled across a stainless steel kiss roll applicator containing the desired silane size solution, and wound directly onto a 12" collet. Alternatively, comparison fibers were made by drawing 204 fibers together in order to asses the influence of glass-to-glass abrasion during drawing. The resulting single-fiber cake was heated for two hours at 110°C to remove water and alcohol generated from the silane. Samples from the resulting glass fibers were analyzed for diameter by SEM and used for micromechanical testing.

The fiber strength distributions were determined using ASTM Standard D3379-75. Single fibers were mounted on cardboard tabs of 1.0", 0.7", 0.6", 0.5" and 0.2" gauge lengths. The fiber tensile strengths were measured using an Instron 1123 at a crosshead speed of 0.02 in. per minute. A minimum of 75 samples were taken at each gauge length. The tensile strength at each gauge length was determined by first fitting test results to a Weibull distribution:

$$G(\sigma) = 1 - e(-l(\tfrac{\sigma}{\beta})^{\alpha})$$

where G is the cumulative probability of failure, l is the gauge length, σ is the applied stress and α and β are Weibull parameters that are characteristic of the glass fiber. With the Weibull parameters defined, the tensile strength for a specific gauge length is calculated from the equation:

$$\sigma(l) = (\beta / l^{(1/\alpha)})\Gamma(\alpha + 1/\alpha)$$

where Γ is the Gamma function. Once calculated, the mean fiber strengths were used to determine the "double box" parameters[2].

Single fibers were mounted on silicone rubber molds and embedded in a resin matrix consisting of 1.0 part Epon® 828 and 1.1 parts Versamide® 140. The resin was cured at 110°C for 1 hour and postcured at 140°C for 45 minutes. The resulting composites were cooled to room temperature and removed from the mold. The composites were mounted on a servohydraulic mechanical testing system fitted with an environmental chamber covering the clamps. The chamber contained viewing windows which allowed sample observation through a stereoscope equipped with crosspolar capability. The environmental chamber was heated to the desired temperature and the composites were pulled under tension at a rate of 0.02 in. per minute until fragmentation of the glass fibers was complete. The lengths of the resulting fragments were measured using a micrometer eyepiece. A minimum of fifty fragments for each composite were required for statistical analysis. The results were fitted to a Weibull distribution. The critical lengths were then calculated from the mean fragment length defined by the Weibull distribution.

3 RESULTS AND DISCUSSION

As glass is drawn from the melt at approximately 1200°C, it rapidly cools from a free flowing liquid to a glass. In the process a large number of highly strained Si-O-Si bonds are incorporated into the glass surface. These are very reactive sites, particularly toward compounds such as water that contain OH groups. When the glass encounters a water spray or an aqueous size bath at the applicator roller, the siloxane bond reacts to form SiOH sites on the surface of the glass[7]. These SiOH sites are then available for reaction with hydrolyzed silanes in the size solution. The attachment of the silane to the glass provides mechanisms for tying the glass to an organic matrix and for preventing additional water molecules from further reacting with the glass to etch the surface.

When glass fibers are abraded, or otherwise damaged, stress is focussed on the tip of the flaw line. In a reaction similar to the initial SiOH formation from strained Si-O-Si bonds on the glass surface, water reacts with the strained Si-O-Si bonds at the tip of the flaw line to further propagate the crack and form new SiOH sites[8]. When available, silanes will bond with these sites and, if the gap is of the appropriate width, heal the flaw[9]. As water sized glass is allowed to age in ambient conditions, more and more surface flaws and active SiOH sites are generated through chemical etching. Previous work[10] has shown that aged, water sized glass fibers commonly used in silane studies have considerably lower tensile strengths as a result of the aging process. Since the new flaws mask the original glass fiber strength and the additional SiOH sites alter the bonding pattern of

Chemically Modified Surfaces

silanes, testing of silane performance and properties should be done on the freshest glass possible in order to reflect actual production conditions.

Table 1 shows the tensile strength for freely drawn glass fibers sized with a series of aminoalkyl silanes. The paired sets consist of a sample drawn as a single fiber and one prepared as a multiple fiber that was cured and then analyzed by isolating a single fiber. The first two sets in the series are dimethoxy silanes containing only two active silyl sites capable of bonding to the glass surface or of forming a siloxane network in the interphase. 3-APMDS contains a short chain aminopropyl functional group and is thus an analogue of A-1100. 4-ADBMDS contains a dimethyl substituted butyl chain bearing the amino group. It is slightly bulkier than 3-APMDS and might be expected to form a more open, less ordered siloxane interphase region.

The second grouping consists of trialkoxy silanes which have three active silyl sites available for both bonding to the surface and for siloxane network formation. A-1100 is an industry standard silane containing a short chain aminopropyl functional group. In comparison, 11-AUTMS contains a much longer 11-aminoundecyl functional group. The third category has only one silane, A-1170, which is a secondary alkylamine bearing two trialkoxy silyl groups. Thus the series in Table 1 explores the change in the number of bonding sites on silicon as well as changes in the character of the amine groups.

Silane	# of Filaments	Tensile Strength, psi				
		1.0 inch	0.7 inch	0.6 inch	0.5 inch	0.2 inch
Dialkoxy Silanes						
4-ADBMDS-s	1	2.97×10^5		3.13×10^5		3.49×10^5
4-ADBMDS-m	204	1.07×10^5		1.78×10^5		2.13×10^5
3-APMDS-s	1	2.11×10^5		2.28×10^5		3.10×10^5
3-APMDS-m	204	2.39×10^5		2.46×10^5		2.65×10^5
Trialkoxy Silanes						
A-1100-s	1	3.57×10^5	3.89×10^5		4.12×10^5	
11-AUTMS-s	1	3.17×10^5	3.38×10^5		3.79×10^5	
Hexaalkoxy Silanes						
A-1170-s	1	4.15×10^5	4.74×10^5		6.25×10^5	
A-1170-m	204	2.18×10^5		1.89×10^5		2.34×10^5

Table 1 Tensile strengths, $\sigma(l)$ of E-glass fibers coated with OSi Specialties Organofunctional Silanes A-1100, A-1170, 11-AUTMS, 4-ADBMDS , and 3-APMDS at several gauge lengths.

Table 1 and Figure 1 show the tensile strength measurements for the single fiber tests (solid curves). The results show a continuous increase in tensile strength as the number of reactive alkoxysilyl groups increases per coupling agent molecule. Thus the dialkoxysilyl 3-APMDS and 4-ADBMDS have a lower tensile strength than the trialkoxysilyl silanes such as A-1100 and 11-AUTMS. Continuing in the sequence, the A-1170 again shows an improvement in tensile strength with increased alkoxysilyl content. These results are consistent with the ability of the three classes of silanes to both bond with the glass surface and form a siloxane network at the interface. The difunctional silane can either form two links with the surface, or form a single link to the surface and condense with one other silane to start a siloxane chain. Further condensation of additional silane molecules can create an extended chain, but never with more than two silyl groups available for attachment to the glass surface. These siloxane chains may have an advantage in projecting into the interphase region to more securely bind or entangle with a matrix resin. However, they also have a distinct disadvantage in trying to protect the glass fiber. If the relatively inert portion of the siloxane chain covers active Si-OH sites or strained Si-O-Si sites on the glass surface, then these sites will be susceptible to later degradation and loss of tensile strength. Further, short chain silanols may readily self condense to form small, very stable cyclic structures held near the surface only through hydrogen bonding.

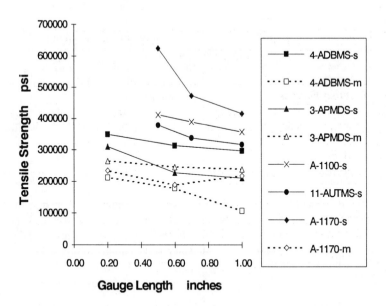

Tensile Strength for Silane Sized Fibers

Figure 1 Tensile strength vs gauge length for E-glass fibers sized with OSi Specialties Organofunctional Silanes A-1100, A-1170, 11-AUTMS, 4-ADBMDS, and 3-APMDS.

The situation is dramatically different for the trialkoxy silanes. The three reactive alkoxysilyl groups allow for a point of attachment and condensation with two additional silyl groups. The result is a long linear or branched siloxane chain with a potential point of attachment at every non-branching silyl group. Continuing this trend, A-1170 can form extensive two and three dimensional networks with multiple bonding sites at any one silyl group. This network formation should make it easier for the A-1170 silane to bridge surface flaws created during glass fiber drawing and prevent them from becoming significantly more severe.

Within each category, there are performance differences that are related to the alkylamine structure. Thus fibers treated with A-1100 have a higher tensile strength than those sized with 11-AUTMS. The aminopropyl group of A-1100 is available to speed the reaction of the silane with glass surface either via direct catalysis of the silyl group or through a more facile absorption/attachment mechanism[11]. Similarly, the 4-ADBMDS contains a primary amine on a substituted butyl chain. This type of configuration is known to readily form a cyclic silyl amine that is very reactive with SiOH groups as well as other active -OH bearing compounds. This method of autocatalysis would cause more rapid reaction with the glass surface and might account for the improved performance of 4-ADBMDS over 3-APMDS.

The tensile strength measurement for glass fibers is not as much a measurement of continuous glass strength as it is a measure of the number and severity of flaws in the glass. Rosen[12] has prepared a "weak link" model for mathematically considering the distribution of flaws on the glass fibers. Thus the glass fiber is treated as a series of small links fused together to form a filament. Using tensile strength measurements at three different gauge lengths, the glass fiber can be described in terms of the fraction with minor flaws and the fraction with severe flaws. The resulting dual flaw populations is considered a "double box" distribution. Based on this concept, Elbirli[13] showed that the populations are continuous and has derived an equation for the average fiber strength, as a function of length:

$$\sigma = \sigma_1 + (\sigma_2 - \sigma_1)[1 - (1-p)^{n+1}]/p(n+1) + (\sigma_4 - \sigma_2)(1-p)^n/n + 1$$

where σ_1 represents the lower limiting strength of a continuous fiber, σ_2 is the lower limiting strength of a continuous fiber not containing a severe surface flaw, and σ_4 is the upper limit for a continuous flaw-free fiber. The fraction of links containing severe surface flaws is represented as p, and n is the number of unit links at a specific gauge length defined as:

$$n = L/l_o$$

where L is the gauge length and l_o is the chosen unit link length. If σ_4 is assigned a value, normally assumed to be 1.5×10^6 psi[13], then three experimental points will allow the solution for σ_1, σ_2 and p at any given unit link length, l_o.

Table 2 shows the calculated σ_1, σ_2 and p values for each of the silanes. In general, the results follow the expected pattern of σ_1 around 1.5 x 10⁵ psi and σ_2 of around 4.0 x 10⁵ psi. The interesting exceptions are the σ_2 values of 1.5 x 10⁶ psi for A-1170 and 11-AUTMS sized fibers drawn as single filaments. Since 1.5 x 10⁶ is also the value assigned to σ_4, the maximum glass fiber strength, these results indicate that a bimodal or "double box" distribution of severe and minor flaws might not be appropriate. Instead, with the right silane chemistry, glass fibers can be drawn with a single continuous flaw population that avoids the most severe flaws.

Table 2 also shows the p value for each of the silanes. The flaw frequency for A-1170 and 11-AUTMS is higher than for silanes such as A-1100 which has a classic "double box" distribution of flaws. A first assumption might be that the greater the number of flaws, the weaker the fiber should be, and generally this holds. The difference is that for those silanes with a single distribution, the number of flaws is higher because all flaws are measured. The fiber is then stronger because there are significantly fewer severe flaws.

The source of the severe flaws is a matter of conjecture, but certainly one of the most significant sources of stress on the fibers is the glass-to-glass abrasion during drawing. Since these fibers were drawn as single fibers, they encountered no other glass until they reached the collet. This is not the case for normal glass which is drawn as a multiple fiber strand.

Silanes	Filaments Drawn	Double Box Parameters		
Dialkoxy		σ_1	σ_2	p
4-ADBMDS-s	1	2.66x10⁵	3.80x10⁵	2.38x10⁻³
4-ADBMDS-m	204	9.70x10⁴	4.00x10⁵	5.00x10⁻²
3-APMDS-s	1	1.85x10⁵	6.31x10⁵	1.10x10⁻²
3-APMDS-m	204	2.26x10⁵	2.81x10⁵	2.79x10⁻³
Trialkoxy				
A-1100-s	1	1.66x10⁵	4.89x10⁵	7.80x10⁻⁴
11-AUTMS-s	1	2.65x10⁵	1.47x10⁶	1.60x10⁻²
Hexaalkoxy				
A-1170-s	1	1.66x10⁵	1.50x10⁶	3.80x10⁻³

Table 2   Calculated tensile strengths parameters from Elbirli's equation. σ_1 is the lower limiting strength value of a continuous fiber, σ_2 is the lower limiting strength value of a continuous fiber not containing a severe surface flaw and p is the fraction of the link population containing a severe surface flaw.

Table 1 and Figure 1 show the tensile strengths for several silane-sized glass fibers pulled as a 204 filament strand (dotted curves). Viewed as sets, the tensile strengths for the multiple filament strands are lower reflecting the additional damage inflicted by the glass-to-glass abrasion as the fibers are collected below the applicator, passed over a winder and wound around the collet. This causes a change from a single population of less severe flaws to a bimodal population including very severe flaws such in the case of A-1170.

Measurements of the critical aspect ratio and the interfacial shear stress for single glass fibers embedded in epoxy resin reveal a different structure/performance relationship. Single fiber composites are placed in an environmental chamber and heated to 25, 40, 50, 60 or 75°C. After applying tension on a composite until the glass fiber no longer breaks with elongation, the fragments are measured and fit to a Weibull distribution. From this the median fragment length (l_f) is determined and the critical length (l_c) is calculated using the formula:

$$l_c = 4l_f/3$$

The critical aspect ratio is then defined as l_c/d, where d is the glass fiber diameter, and can be calculated as a quick method for comparing glass fiber performance. The critical aspect ratio reflects the ability of the interface to transfer sufficient stress to the glass surface to break the fiber, but does not consider the tensile strength of the glass fiber. For comparable glass fibers, lower l_c/d values indicate better size performance.

Table 3 and Figure 2 show the critical aspect ratios for each silane-sized fiber embedded in an epoxy composite at 25, 50 and 75°C. The results illustrate the requirement to compare the critical aspect ratios only for evaluating trends for fibers of the same tensile strength. In this case the dialkoxy silanes have a relatively lower (better) critical aspect ratios than the tri- and hexaalkoxy silanes. This is in part due to the much lower tensile strengths of the glass fibers sized with the dialkoxy silanes. Obviously it is easier to break weaker glass fibers into smaller pieces. The fibers do show a predictable increase in the critical aspect ratio as the temperature increases. Further, beyond the T_g of the epoxy matrix (55°C), the rate of increase accelerates rapidly.

The interfacial shear stress measures the ability of the organofunctional silane to transmit stress from the matrix to the glass fiber. Higher values reflect a more effective interface and thus better silane performance. Interfacial shear stress (τ_e) is calculated from the tensile strength of the fiber at its critical length (σ_{l_c}) and the critical length of the embedded glass fiber (l_c) according to the equation:

$$\tau_e = \sigma_{l_c}d/2l_c$$

Table 3 and Figures 4 and 5 show the interfacial shear stress transmission (ISST) for glass fibers sized with each of the silanes and tested over a temperature range from 25 to 75°C. The results at 25°C demonstrate that the ISST values can vary considerably as the structure of the silane is changed. In contrast to the results

Silane	Filaments Drawn		Temperature		
			25°C	50°C	75°C
Dialkoxy					
4-ADBMDS-s	1	ISST	5.510×10^3	2.817×10^3	8.585×10^2
		lc	0.43	0.67	2.4
		lc/d	38.4	59.8	214.3
4-ADBMDS-m	204	ISST	5.470×10^3	3.355×10^3	4.505×10^2
		lc	0.37	0.6	2.21
		lc/d	26.6	43.2	159
3-APMDS-s	1	ISST	8.030×10^3	5.912×10^3	1.078×10^3
		lc	0.42	0.55	2.19
		lc/d	37.2	48.7	193.8
3-APMDS-m	204	ISST	7.398×10^3	3.043×10^3	8.050×10^2
		lc	0.31	0.65	1.97
		lc/d	31	65	197
Trialkoxy					
A-1100-s	1	ISST	2.202×10^3	1.385×10^3	4.780×10^2
		lc	1.2	1.87	5.11
		lc/d	112.2	174.8	477.6
11-AUTMS-s	1	ISST	8.230×10^3	3.333×10^3	1.239×10^3
		lc	0.79	1.56	3.07
		lc/d	68.9	136.8	269
Hexaalkoxy					
A-1170-s	1	ISST	1.503×10^4	5.665×10^3	1.900×10^3
		lc	0.54	1.33	3.33
		lc/d	47.4	116.7	292.1

<u>Table 3</u> Critical lengths, l_c of single and multiple E-glass fiber(s) coated with either OSi Specialties Organofunctional Silanes 4-ADBMDS, 3-APMDS, A-1100, 11-AUTMS, and A-1170. ISST is the interfacial shear stress, and l_c/d is the critical aspect ratio.

for tensile strengths, the ISST values for the dialkoxysilane treated fibers are interspersed with the trialkoxy silane treated fibers. This indicates that the relative ability of each silane to attach to the surface is not the determining factor. Rather, the differences are the result of the interaction at the interphase/matrix boundary.

Critical Aspect Ratio

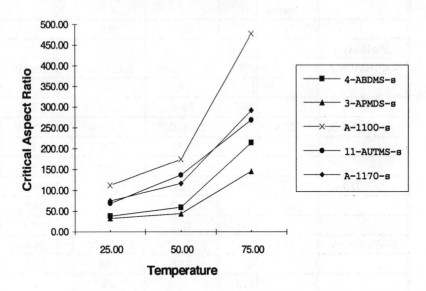

Figure 2 Critical aspect ratio vs sample temperature for single fiber expoxy composites made from E-glass fibers sized with OSi Organofunctional Silanes A-1100, A-1170, 11-AUTMS, 4-ABDMS, and 3-APDMS.

The interfacial shear stress transmission will improve as the interphase/matrix region is strengthened. All of the silanes tested in this paper are amino functional silanes capable of bonding with the epoxy matrix to give a base strength to this region. Further improvement can come from increased entanglement of the matrix in the interphase which can be achieved by improving the solubility of the matrix resin in the interphase. During curing of the sizing, the dialkoxy silanes condense to form straight chain and cyclic siloxanes, with many more molecules tied into the interphase by hydrogen bonding rather than covalent Si-O-Si bonds. The result is an interphase which is more easily penetrated by the epoxy resin.

As the number of alkoxy groups on the silicon is increased from two to three, it is possible to form extensive, highly crosslinked siloxane networks during curing. A-1100, which has three ethoxy groups attached to silicon, condenses to form layers of tight protective sheathing around the glass fiber. This is much more difficult for the matrix resin to penetrate and react with. Thus the ISST is lower even though the glass fibers have higher tensile strengths. 11-AUTMS contains a long alkyl chain that disrupts the formation of a tight network around the fiber. The matrix resin is better able to penetrate the silane interphase and raise the ISST.

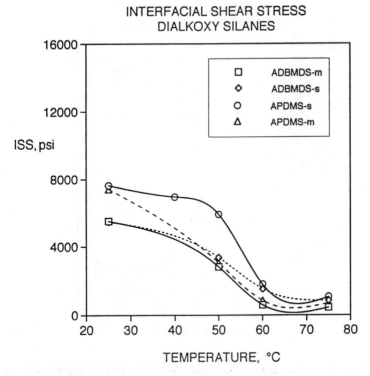

Figure 3 Interfacial shear stress transmission v. temperature for single fiber epoxy composites made from E-glass fibers sized with dialkoxy silanes.

Similarly, A-1170 should have better matrix resin penetration as a result of being bulky enough that tightly fitted sheets of siloxane may not form around the fiber. The interfacial shear stress transmission can also be improved by increasing interphase/matrix crosslink density to stiffen the interphase. Since A-1170 has two trialkoxysilyl groups, it can easily create a much more heavily crosslinked siloxane network which will extend from the glass surface out into the interphase. For example, if only one of the silyl groups of the A-1170 bonds directly to the glass surface, then the other is available to bond into the multiple silane layers around the glasss fiber to form an extended siloxane network. This not only creates a more highly crosslinked, more rigid interphase, it also presents more reactive amine sites per molecule for bonding with the epoxy matrix.

With all of the silanes, the interfacial shear stress transmission decreases as the temperature increases. It is unlikely, particularly in the absence of water, that there is any significant chemical degradation at the glass fiber/interphase boundary over the temperature range being considered here. The most dramatic drop in ISST occurs as the temperature is raised through the 50° to 60°C range, which coincides with the T_g (55°C) of the epoxy resin as measured by DSC. Thus the major loss in the measured ISST is the result of loss of strength in the matrix, not the interphase.

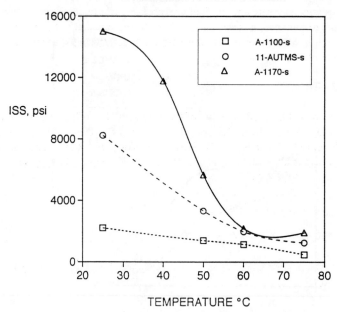

**INTERFACIAL SHEAR STRESS
TRI&HEXA-ALKOXY SILANES**

Figure 4 Interfacial shear stress transmission vs. temperature for single fiber epoxy composites made from E-glass fibers sized with tri and hexaalkoxy silanes.

The continued integrity of the interphase is demonstrated by the constant ordering of the silane ISST's. Thus for example, A-1170 has a significantly higher ISST than A-1100 over the entire temperature range. This is consistent with the concept that the siloxane/epoxy resin interphase should not be as susceptible to temperature changes as it is more highly crosslinked and has a higher T_g than the epoxy matrix.

4 CONCLUSION

The tensile strength and flaw distributions of glass fibers drawn with aminoalkylsilanes are dependent on the chemical structure of the chosen silanes. As the number of alkoxy groups attached to silicon increases, the bonding to the glass surface and siloxane network formation at the interfacial layer increases. In turn, this better protects the glass from the most severe surface flaws and increases the tensile strength of the fibers.

Similarly, the performance of glass fibers in reinforcing epoxy composites are dependent on the choice of the structures of the aminoalklsilanes employed as a sizing. The interphase/matrix region defined by the silane is the key interaction in relaying stress from the matrix to the glass fibers. The altering of the structure of the silanes increases the penetration and entanglement of the matrix with the siloxane interphase and thus increases the interfacial shear stress transmission. The interfacial shear stress transmission can also be improved by increasing the crosslink density of the siloxane network to create a more rigid interphase/matrix region having a higher T_g. As the temperature increases through the T_g of the epoxy matrix, the matrix becomes the part of the composite with the least ability to transmit stress while the interphase maintains its integrity.

5 REFERENCES

1. a) S. Sterman and H.B. Bradley, Proc. of the 16th Ann. Tech Conf. Rein. Plast , SPI, Session 8-D, 1961. b) M.E. Schrader, I. Lerner and F.J. Doria., Mod. Plast , 1965, 45, 195. c) O.K. Johannson, F.O. Stark, G.E. Vogel and R.M. Fleischmann., J. Compos. Mater., 1967, 1, 278. d) H. Ishida and J.L. Koenig , J. Polym. Sci., Polym Phys. Ed., 1980, 18, 1931.

2. D.L. Marquart, J. Soc. Ind. Appl. Math., 1963, 2, 431.

3. P.S. Chua and M.R. Piggott, Comp. Sci. Tech., 1985, 22, pp 33, 107, 185, 245.

4. A.N. Netravali, D. Stone and T.T.L. Topolesk, Comp. Sci. Tech., 1989, 34, 289.

5. U. Gaur and B. Miller, Comp. Sci. Tech., 1989, 34, 35.

6. W.A. Fraser, F.H. Ancker, A.T. DiBenedetto and B. Elbirli, Polym. Copmos., 1983, 4, 238.

7. B.C. Bunker, D.M. Haaland, T.A. Michalske and W.L. Smith, Surf. Sci., 1989, 22, 95.

8. T.A. Michalske and B.C. Bunker, J. Appl. Phys., 1984, 56, 2686.

9. D.L. Vaughan and J.W. Sanders, Proc of the 29th Ann. Conf., SPI, Comp. Inst., Session 13-A, 1974.

10. J.A. Gómez and J.A. Kilgour, Proc. of the 48th Ann. Conf., SPI, Comp. Inst., Session 9-D, 1993.

11. C.H. Chiang, H. Ishida and J.L. Koenig, J. Colloid Interface Sci., 1980, 74, 396.

12. B.W. Rosen, 'Fiber Composite Materials', ASM, Metal Park, Ohio, Chapter 3, 1965.

13. B. Elbirli, Ph.D. Dissertation, The University of Connecticut, 1979.

Stability and Reactivity of Dimethylethoxysilane

Richard E. Johnson and Douglas I. Ford
DEPARTMENT OF CHEMISTRY, LETOURNEAU UNIVERSITY,
LONGVIEW, TX 75607, USA

1 INTRODUCTION

In this paper, the chemistry of the compound dimethylethoxysilane (DMES) is discussed especially as it relates
to waterproofing silica surfaces. Some of the desirable
properties of this compound are that it readily reacts
with silica in the vapor phase, it is a low boiling point
liquid (54°C) and the by-product of its reaction with
silica is the rather inert substance, ethanol. It is
currently used by NASA to re-waterproof the HRSI shuttle
tiles before relaunching the vehicle.

Very little information is available on this particular compound in the literature or even on related silane
compounds that have both a hydride group and an alkoxy
group. Since the close proximity of two groups often
drastically affects the chemical behavior of each group,
chemical reactions were carried out in the laboratory
with DMES to verify the expected behavior of these two
functional groups located on DMES. Some of the reactions
tested would be potentially useful for quantitative or
qualitative measurements on DMES. To study the reactions
of DMES with silica surfaces, cabosil was used as a
silica substrate because of its high surface area and the
ease of detection by infrared spectroscopy as well as
other techniques.

2 CHEMICAL REACTIONS OF DIMETHYLETHOXYSILANE (DMES)

DMES has the following structure and physical properties.

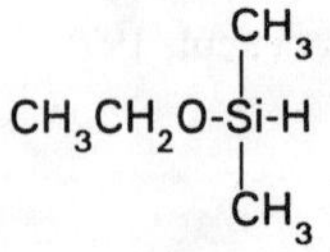

$$\begin{array}{c} CH_3 \\ | \\ CH_3CH_2O\text{-}Si\text{-}H \\ | \\ CH_3 \end{array}$$

Density 0.751 g/ml at 25°C

Boiling point 54°C

Refractive index 1.365

The methyl-silicon bonds are chemically inert toward ordinary chemical reagents (acids, bases, water, oxidizing and reducing agents). Most of the chemistry of this compound is due to the ethoxy and hydride groups. In contrast to carbon-ethoxy linkages, the silicon-ethoxy bond is readily hydrolyzed by water. The hydrogen of the Si-H linkage is a fairly strong reducing agent, whereas the C-H linkage is quite inactive as a reducing agent. This is probably due to the smaller electro-negativity of the silicon atom compared to carbon. DMES, then, has two fairly aggressive functional groups. These two groups account for most of its chemical reactions at ordinary temperatures and ordinary chemical environments.

In the following section, reactions of DMES which were investigated are discussed. Various applications of each reaction are also given. Figures 1 and 2 summarize these reactions.

Water Hydrolysis

When water is added to DMES, separation in two layers occurs due to immiscibility. Agitation of the liquids for a few minutes produces a homogeneous solution. The obvious hydrolysis reaction would be

$$H_2O + H\text{-}\underset{\underset{CH_3}{|}}{\overset{\overset{CH_3}{|}}{Si}}\text{-}OEt \longrightarrow H\text{-}\underset{\underset{CH_3}{|}}{\overset{\overset{CH_3}{|}}{Si}}\text{-}OH + EtOH \qquad (1)$$

However, the resulting silanol has not been isolated[1]. Indeed, the combination H-Si-OH does not occur in any known stable compound and is at best a very reactive intermediate. It is in this respect that DMES is unique compared to other common silylating agents. GCMS confirmed that the reaction of DMES and water produced ethanol and 1,1,2,2-tetramethylsiloxane. The net reaction is therefore

$$2\,H\text{-}\underset{\underset{CH_3}{|}}{\overset{\overset{CH_3}{|}}{Si}}\text{-}OEt + H_2O \longrightarrow H\text{-}\underset{\underset{CH_3}{|}}{\overset{\overset{CH_3}{|}}{Si}}\text{-}O\text{-}\underset{\underset{CH_3}{|}}{\overset{\overset{CH_3}{|}}{Si}}\text{-}H + 2EtOH \qquad (2)$$

When DMES is applied to wet silica, this reaction will compete with the reaction of the DMES with silanol groups. However, infrared studies indicate that this is not a major factor for silica surfaces with physically adsorbed water, but could certainly be a major reaction if liquid water is present.

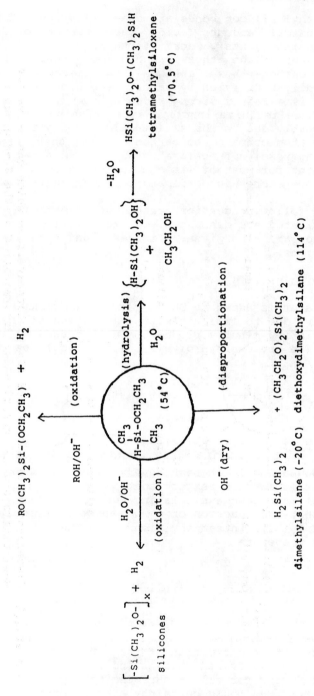

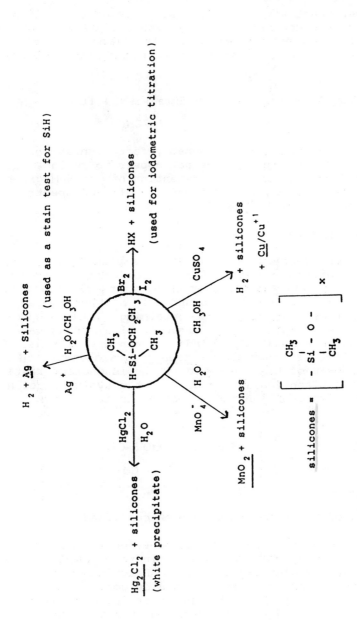

Figure 2 Miscellaneous oxidations of dimethylethoxysilane which can serve for qualitative and for quantitative determination of the Si-H group.

Reaction With Silica-OH Groups

Silylating agents typically hydrolyze to silanols which then react with the fixed silanols of the silica surface to produce the Si-O-Si (siloxane) linkage. However, since DMES does not produce a stable silanol, it is likely that it reacts directly with the surface hydroxyls:

$$
\text{Silica-OH} + \text{EtO-}\underset{\underset{CH_3}{|}}{\overset{\overset{CH_3}{|}}{Si}}\text{-H} \longrightarrow \text{Silica-O-}\underset{\underset{CH_3}{|}}{\overset{\overset{CH_3}{|}}{Si}}\text{-H} + \text{EtOH} \qquad (3)
$$

The reaction occurs reasonably fast even at room temperature, although it does not appear to replace all of the hydroxyl groups. The retention of the Si-H group during this reaction is confirmed by infrared spectra of cabosil treated with DMES (Figures 3, 4).

Disproportionation

A major impurity in "old" DMES samples is found by GCMS to be diethoxydimethylsilane. For example, a bottle of DMES which was about two or three years old contained about 5% of the diethoxy compound. It has been noted in the literature that compounds containing H-Si-OEt disproportionate especially in the presence of a strong base which acts as a catalyst[2]. Indeed, when dry NaOH is added to DMES, a vigorous reaction occurs accompanied by a small amount of heat released (which is characteristic of disproportionation) and also with the evolution of a gas. The disproportionation reaction is as follows.

$$
2 \text{ H-}\underset{\underset{CH_3}{|}}{\overset{\overset{CH_3}{|}}{Si}}\text{-OEt} \xrightarrow[\text{(dry)}]{OH^-} \text{H-}\underset{\underset{CH_3}{|}}{\overset{\overset{CH_3}{|}}{Si}}\text{-H} + \text{EtO-}\underset{\underset{CH_3}{|}}{\overset{\overset{CH_3}{|}}{Si}}\text{-OEt} \qquad (4)
$$

(b.pt. 54°C)   (b.pt. -20°C)   (b.pt. 114°C)

Both of the products have been confirmed by GCMS.

The above reaction might account for the slow pressure build up observed in containers of DMES and the presence of the diethoxy impurity. Reactions 5 and 6 (discussed in next section) could also contribute H_2 gas as well. Figure 5 shows this pressure increase over a period of several days. The non-congruency of the lines of Figure 5 is likely due to the absence of light during the night hours of the experiment indicating a sensitivity of the reaction to light. Decomposition did continue to occur in an amber polypropylene bottle as well, indicating that some reaction occurs without catalytic effects of the glass surface or exposure to light.

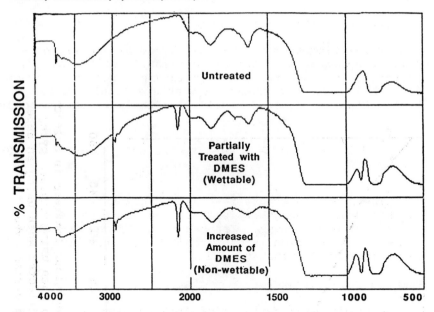

Figure 3   Infrared spectra of cabosil powder in various
stages of reaction with dimethylethoxysilane
vapors at room temperature.

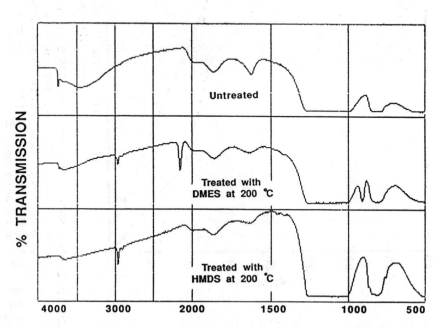

Figure 4   Infrared spectra of cabosil powder before and
after treatment with dimethylethoxysilane
(DMES) and hexamethyldisilazane (HMDS) vapors.

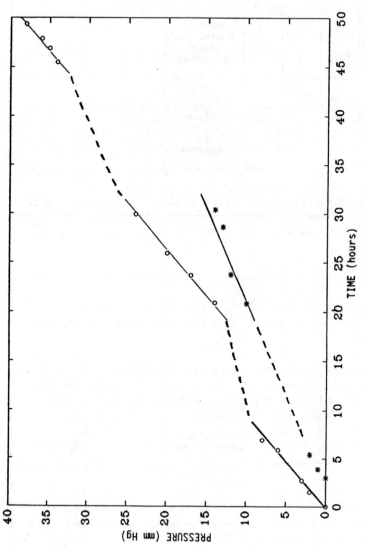

<u>Figure 5</u>   Pressure build-up over DMES in a closed container immersed in a 26°C water bath. Circles denote data collected when using a clear pyrex glass container. Asterisks denote data collected when using an amber polypropylene container. Dashed lines are periods when the lab lights were off.

The Si-H Group as a Reducing Agent

Hydrides of silicon are known to be good reducing agents. The -1 oxidation state of the hydrogen can be increased to 0 to produce H_2 gas or to +1 to produce water or hydrogen ions. Reactions which were examined involving the oxidation of DMES are given below.

Base Catalyzed Oxidation with Alcohols.

$$ROH + H\underset{\underset{CH_3}{|}}{\overset{\overset{CH_3}{|}}{Si}}OEt \xrightarrow{OH^-} RO\underset{\underset{CH_3}{|}}{\overset{\overset{CH_3}{|}}{Si}}OEt + H_2 \tag{5}$$

This reaction was utilized to synthesize diethoxy-dimethylsilane in order to verify its peak as a decomposition product in the GC of DMES. Ethanol was used as the alcohol and the reaction was vigorous even at room temperature.

Base Catalyzed Oxidation with Water.

$$n\,H_2O + n\,H\underset{\underset{CH_3}{|}}{\overset{\overset{CH_3}{|}}{Si}}OEt \xrightarrow{OH^-} (\underset{\underset{CH_3}{|}}{\overset{\overset{CH_3}{|}}{Si}}O)_n + n\,H_2 + n\,EtOH \tag{6}$$

(silicones)

Alkaline water reacts with both the hydride and ethoxy groups to give a bifunctional monomer which can then form silicone polymers.

Oxidation with Hg^{+2}(aq). An aqueous solution of mercuric chloride is rapidly reduced by DMES to produce mercurous chloride and silicones. The combination of an oxidizing agent and water with DMES can always be expected to produce silicones. This reaction can be used to detect the Si-H group in silanes by the observation of the white precipitate, Hg_2Cl_2.

Oxidation with Aqueous I_2 or Br_2. Both I_2 and Br_2 oxidize DMES to produce silicones and either HI or HBr. The reaction with I_2, as expected, is slower than with Br_2. Both reactions have been used to perform quantitative oxidation-reduction titrations of DMES. Excess iodine or bromine was allowed to oxidize the DMES and then the excess halogen was determined using standardized sodium thiosulfate. The excess bromine was generated by using potassium bromate and excess potassium iodide.

Oxidation with Cu^{+2}. When a saturated solution of $CuSO_4$ in methanol comes in contact with DMES in liquid or vapor form, it rapidly deposits a brown solid and produces hydrogen gas. The solid is probably metallic copper and Cu^+ salts. It provides a very sensitive test for DMES vapors.

Oxidation with Ag^+. A silver nitrate solution (0.1 M in 50/50 water-methanol) reacts rapidly with DMES to produce H_2 and metallic silver. Initially, a yellow color appears which then disappears or is masked by the gray to black silver precipitate. This was used to examine silica shuttle tiles which had been treated with DMES. When the tile was treated with DMES, spraying it with the silver nitrate solution produced a gray color, presumably due to the reduction of the silver ions by the Si-H.

3   GAS CHROMATOGRAPHY STUDIES

Gas chromatograms were obtained for DMES in various stages of reaction with water and with sodium hydroxide. Dimethylsilicone stationary phases were employed on both packed and capillary columns in the chromatograph. The packed columns were operated isothermally at 50°C and utilized a flame ionization detector. The capillary columns utilized temperature programming and a mass spectrometer for the detector. All of the major peaks and many of the minor peaks were identified from the mass spectra obtained. Nearly all of the peaks could be accounted for in terms of expected hydrolysis and dispro-portionation products (reactions 2 and 4), with the disproportionation products then reacting with incidental moisture and the parent DMES molecule.

The following compounds are identified as major peaks coming off before twenty minutes on the capillary column (eight minutes on the packed column):

$$(CH_3)_2Si(OCH_2CH_3)_2 , (CH_3)_2HSiOSiH(CH_3)_2 , CH_3CH_2OH , (CH_3)_2H_2Si ,$$

$$(CH_3CH_2O)Si(CH_3)_2O(CH_3)_2SiH , (CH_3)_2HSiOSi(CH_3)_2OSiH(CH_3)_2 .$$

Many of the minor components were identified as being polysiloxanes, both linear and cyclic. The above mentioned components were also found in a DMES sample which was two to three years old.

4   INFRARED STUDIES OF DMES TREATED SILICA

Cabosil is a high surface area silica powder which is chemically similar to crystalline quartz and fibrous silica. Cabosil samples can be prepared for infrared spectroscopy studies by lightly pressing the powder sample between two salt plates of a demountable liquid cell. The apparatus used for room temperature studies of cabosil and DMES is shown in Figure 6. After reacting in the apparatus, a portion of the cabosil powder was then pressed between the salt plates to a thickness of about 0.5 mm. Before discussing the results of infrared studies, a brief discussion of silica surfaces is given.

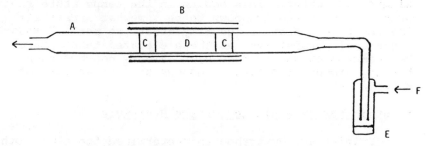

Figure 6   Apparatus for chemical treatment of cabosil powder. Legend: A - quartz tube, B - furnace, C - porous silica plugs, D - cabosil sample, E - DMES or other agents, F - nitrogen gas.

The Nature of Silica Surfaces

Silica is known to have a strong affinity toward water adsorption. This tendency is due to the silica surface hydroxyls of which there are two types, isolated hydroxyls and vicinal hydroxyls[4]. The vicinal hydroxyls can form hydrogen bonds to each other and can retain water on the surface by forming hydrogen bonds to the adsorbed water. Surprisingly, it appears that isolated hydroxyls do not hydrogen bond to water molecules (no infrared shift) and therefore contribute little to water adsorption[4]. However, Lewis base compounds such as amines are preferentially adsorbed on the isolated hydroxyls. Also, other factors being equal, the isolated hydroxyls are more reactive than the hydrogen bonding hydroxyls[4-6].

Figure 3 shows the infrared spectrum of cabosil obtained by pressing the cabosil powder between two salt plates 0.5 mm apart. This sample technique was used previously and the absorption band assignments are as follows[3].

3750 cm^{-1} is the stretching vibration of isolated surface hydroxyls (sharp peak due to no interaction of isolated hydroxyl groups)

3660 cm^{-1} is due to vicinal hydroxyl groups which hydrogen bond to each other and thus give a broad absorption band

3450 cm^{-1} is a broad band due to water adsorbed on the silica surface

Typically, silica surfaces have from one to four hydroxyls per 100 square angstroms of surface[7]. Raising the temperature destroys the vicinal hydroxyls by eliminating water. This occurs in the temperature range of 450 - 800°C and is known to be somewhat irreversible. The isolated hydroxyls (usually about one per 100 square angstroms) persist even at these elevated temperatures. Replacement of the surface hydroxyls by non-hydrogen bonding groups drastically reduces the wettability of the material.

A Method for Determining Surface Hydroxyls

Fripiat and Uytterhoeven[7] determined the OH content of cabosil using the methyl-Mg Grignard reagent to generate methane gas with the surface OH groups. Cabosil was shown to contain one to four hydroxyls per 100 square angstroms of surface area depending upon the temperature. This value should be applicable to the shuttle tile surface as well since it is essentially pure silica except that it would have a much smaller surface area per gram than cabosil. The technique developed here to determine surface hydroxyls on silica is as follows.

Substitute Cl for Surface OH Groups[8].

$$\text{Silica-OH} + CCl_4 \xrightarrow[N_2]{550°C} \text{Silica-Cl} + COCl_2 + HCl \qquad (7)$$

Hydrolyze the chloride (or fluoride) with water.

$$\text{Silica-Cl} + HOH \longrightarrow \text{Silica-OH} + HCl \qquad (8)$$

Titrate the Liberated HCl (or HF) with Standardized NaOH Solution.

We were able to replace the hydroxyls of silica using freon 12, CCl_2F_2. The reaction is conjectured to be

$$\text{Silica-OH} + CCl_2F_2 \xrightarrow[550°C]{} \text{Silica-F} + COCl_2 + HF \qquad (9)$$

It was also found that freon 22, $CHClF_2$, worked equally well to produce the fluorinated surface.

Five grams of hydrolyzed cabosil halide neutralized about 35 ml of 0.1 M NaOH. The same amount of shuttle tile treated the same way only neutralized 0.4 ml of the same NaOH solution. From this, it can be calculated that the cabosil contains about 0.7×10^{-3} moles of surface hydroxyls per gram. This is consistent with previous data reported in the literature using a completely different method[7]. The shuttle tile silica by the same calculations then has about two orders of magnitude less hydroxyl groups (due to less surface area) or about 1×10^{-5} moles/gram. From these data, it can be calculated that the stoichiometric amount of DMES needed to completely react with shuttle tiles is about one milligram of DMES per gram of tile. Cabosil is known to have a surface area of about 200 m^2/g and therefore the shuttle tiles silica is estimated to have less than 10 m^2/g silica of surface area. This low hydroxyl content of the shuttle tiles explains why it is difficult to see reacted DMES on treated tiles by infrared spectroscopy but it can be seen on cabosil surfaces.

Figure 7 shows the spectra of cabosil samples before and after treatment with CCl_4 and CCl_2F_2 (freon 12) at 550°C. Both reagents cleanly remove the surface hydroxyls and replace them with chlorine or fluorine atoms. Both chlorinated and fluorinated cabosil surfaces give essentially the same spectra, except for the weakly discernible SiF band at 900 cm^{-1}. The surface treated with CCl_4 was shown to be chlorinated by hydrolysis of the surface and treating the solution with $AgNO_3$ to obtain the white AgCl precipitate. The surface treated with freon 12 is likely to be fluorinated, since a negative test for chloride was obtained for the hydrolyzed product, and a band at 900cm^{-1} is observed in freon-treated samples. Interestingly, both fluorinated and chlorinated cabosils and shuttle tile silica resisted wetting with water until hydrolysis took place, which was after about one hour of contact with water. Similar results for fluorinated porous glass were obtained using an entirely different process[9].

Infrared Detection of Attached Silyl Groups

Figure 3 shows the effect of reaction of the cabosil with DMES. As water adsorbency decreases, the hydroxyl peaks also diminish as well. Note, however, a new prominent peak at 2150 cm^{-1} which is due to the Si-H group. This peak persists even after heating the cabosil at 130°C under a vacuum. This indicates that the Si-H group does not significantly react with the surface hydroxyls. The small peaks at about 2900 cm^{-1} are due to methyl groups of the silane that have bonded to the surface.

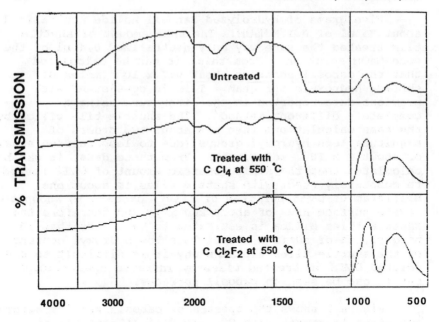

Figure 7   Infrared spectra of cabosil before and
after treatment with carbon tetrachloride
or freon 12.

Figure 4 contrasts cabosil surfaces reacted with
DMES and hexamethyldisilizane (HMDS). The HMDS
provides $(CH_3)_3Si-O-$ groups for bonding to the surface by
replacing the surface hydroxyls. DMES provides
$(CH_3)_2Si(H)O-$ groups for bonding. The differences can be
seen in the infrared spectra. As expected, both HMDS
and DMES treated cabosil give infrared peaks characteristic of C-H groups around 2900 cm^{-1}, but only the DMES
spectrum has the Si-H band at $2150^{-1}cm$.

Thermal Stability of DMES Treated Cabosil

In order to examine the thermal stability of DMES
treated silica, the specially constructed gas cell shown
in Figure 8 was used. A heating tape was wrapped around
the center portion of the cell to control the temperature
in the cell. A thin silica wafer was made by pressing
cabosil powder between two stainless steel plates in a
hydraulic press to a pressure of about 5000 lb/sq in. The
resulting wafer was about 1.5 x 2.0 cm and weighed about
40 to 50 mg. The wafer could then be mounted into the
holder which was then placed into the cell. Generally,
the wafers were dried at room temperature by applying a
vacuum and then reacted with DMES vapor at about 100 mm

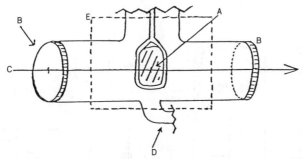

<u>Figure 8</u> Infrared cell used for thermal studies of
DMES-treated cabosil.
Legend: A - silica wafer ($\approx$ 3 cm^2 x 0.25 mm,
$\approx$ 45 mg, 5000 lb/in^2), B - sodium chloride
plates, C - infrared beam, D - access port,
E - heated area.

Hg pressure. The excess DMES was removed by vacuum
again and the IR spectrum obtained. Subsequently, 50%
R.H. air was admitted to the cell and then heated to
various temperatures, held at the temperature for twenty
minutes, and then the air was expelled and the cell
cooled down to room temperature before the IR spectrum
was taken. Representative spectra are given in Figure 9.

The stability of DMES on silica is inferred by
observing the Si-OH, C-H and Si-H peaks. It is evident that
the groups C-H and Si-H begin to diminish in the 400 -
500°C range while the Si-OH peaks begin to increase.
This range for thermal breakdown is consistent with other
work for silylated silica surfaces containing Si-H and Si-
CH groups which also report decomposition of these
groups in the 400 - 500°C range in air[10,11].

After heating to the highest temperature, 640°C, a
silica wafer was retreated with DMES under the same
conditions as previously described. Figure 10 indicates
that there are qualitative differences in the silane
bonded to silica upon retreatment as compared to initial
treatment. The broad Si-OH peak was smaller in the
retreated sample. This can be explained by the known
phenomena of irreversible dehydration of silica surfaces
at high temperatures[4]. Adjacent Si-OH groups on the
surface are converted to Si-O-Si bridges at high
temperatures and these bridges persist even when the
sample is recooled to room temperature.

More significantly, it is noted that even though the
C-H and Si-H peaks reappear on the DMES for the retreated
silica, the Si-H peak is diminished in size, suggesting

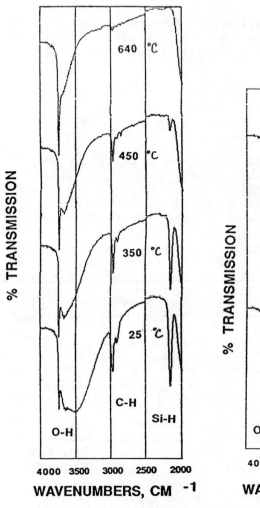

Figure 9   Infrared spectra
           of DMES-treated
           cabosil wafers
           at various
           temperatures.

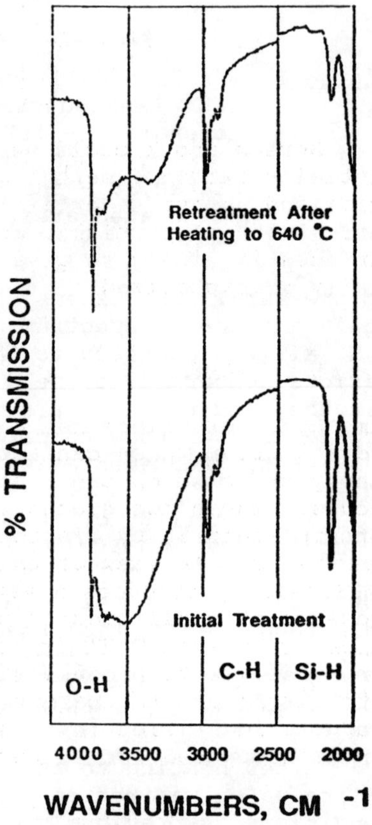

Figure 10 Infrared spectra
           of DMES-treated
           cabosil wafer
           showing the
           difference in
           silyation for a
           new sample which
           had been treated
           and then thermally
           degraded before
           treatment.

that the thermally degraded surface has oxidized some of the Si-H groups of the reacting DMES. Also it is noted that in Figures 9 and 10 the spectra of the samples that have experienced a temperature of 640°C indicate a new kind of surface hydroxyl may be forming. These results are consistent with the following representation of the thermal degradation of the silylated surface.

$$
\begin{array}{c}
X \\
| \\
\text{Silica} -\!\!\text{O}-\!\!\text{Si}-\!\!\text{OH} \\
| \\
Z
\end{array}
$$

Here X and Z could be additional OH groups or partially oxidized methyl groups which could function as oxidizing agents. Certainly it is evident that repeated retreatment of partially degraded silylated silica surfaces is likely to give new surfaces which are not easily characterized.

5   CONCLUSIONS

Dimethylethoxysilane has a varied chemistry involving the hydride group which can act as a reducing agent and the ethoxy group which can react with water and silanol groups. These two groups can switch places intermolecularly by disproportionation under basic conditions. In view of this, it is not surprising that DMES readily produces a variety of products in the presence of small amounts of moisture or bases. A major impurity in aged DMES is diethoxydimethylsilane which can hydrolyze to produce silanols which in turn can react with itself and the parent DMES compound. Most of the products identified in an aged sample of DMES by GCMS can be accounted for in terms of this reaction scheme.

Using a high surface area form of silica called cabosil, infrared spectroscopy studies indicate that DMES reacts with the Si-OH groups by way of the ethoxy group. The silylated surface degrades in air beginning at temperatures of about 400°C which coincidentally is the same temperature the vicinal (adjacent) Si-OH groups of silica begin dehydrating to form Si-O-Si bridges. The infrared studies also suggest that the thermally degraded silylated surfaces which are retreated with DMES will produce attached silyl groups which are varied and not easily characterized. After many retreatments with DMES, the structure of the silylated surface could be much different from the initially treated surface.

ACKNOWLEDGMENTS

This work was supported in part by a grant from NASA Johnson Space Center in Houston, Texas. Mr. Keith Albyn and Mr. Randy Peters are acknowledged for their assistance in some of the GCMS work. Also, several undergraduate students from LeTourneau University assisted in the experimental work.

REFERENCES

1. E.A. Ebsworth, 'Volatile Silicon Compounds', Pergamon Press, 1963.

2. C. Eaborn, 'Organosilicon Compounds'. Academic Press, 1960.

3. R.S. McDonald, J. Am. Chem. Soc., 1957, 79, 850.

4. M.L. Hair, 'Chemically Modified Surfaces, Vol. 1, Silanes, Surfaces and Interfaces', Gordon and Breach Science Publishers, 1986.

5. W. Hertl and M.L. Hair, J. Phys. Chem., 1971, 75, 2181.

6. M.R. Basila, J. Chem. Phys., 1961, 35, 1151.

7. J.J. Fripiat and J. Uytterhoeven, J. Phys. Chem., 1962, 66, 800.

8. J.B. Peri, J. Phys. Chem., 1966, 70, 2937.

9. T.H. Elmer and I.D. Chapman and M.E. Nordberg, J. Phys. Chem., 1963, 67, 2219.

10. J. Mahias and G.J. Wannamacher, J. Colloid Interface Sci., 1988, 1, 125.

11. J.J. Pesek, 'Chemically Modified Oxide Surfaces', vol. III, Gordon and Breach Science Publishers, 1990, 99.

Molecular Dynamics of Liquid Chromatography: Chain and Solvent Structure Visualization

Mark R. Schure

THEORETICAL SEPARATION SCIENCE LABORATORY, ROHM AND
HAAS COMPANY, SPRING HOUSE, PA 19477, USA

Introduction

The majority of applications which utilize chemically-bonded phases in liquid chromatography use either C-18 or C-8 technology. In these cases silica, which has traditionally been used as the support material, is reacted with a silane containing an alkyl chain (either octadecyl or octyl) to form the bonded surface. The silane typically contains two methyl groups or methoxy groups in addition to the long chain alkyl group.

A number of theoretical studies have focused on describing the interaction of the alkyl chains with solvent and solute in attempts to quantitatively clarify what chemical and physical parameters are most dominating in the retention process and in providing an a priori set of equations that attempt to predict retention. These treatments include statistical lattice treatments[1,2] and the solvophobic theory[3] amongst many other models of more or less empirical origin. These models have been the subject of a recent review article concerning the mechanism of retention in reversed-phase liquid chromatography[4]. Only a few studies have attempted to describe a chromatographic retention mechanism at the atomic level of detail because of the extreme complexity of the system and because not all of the atomic detail, e.g. the silica surface structure, is known in detail. For the cases[5,6] where atomic detail has been considered, chiral stationary phases were modeled for specific interactions in the absence of solvent and silica support using computer simulation methods.

The solvent effects which govern liquid chromatography with bonded phases are extremely dominant, as viewed by the effect the solvent composition has on chromatographic retention. Although a wealth of chromatographic experiments have been performed which have attempted to elucidate the structural and dynamical effects of the bonded phase, these measurements can only be used as suggestive information and can not directly lead to structure or mechanism, due to their essentially thermodynamic origin. On the other hand, a number of spectroscopic techniques have been used to study the bonded phase and support conformation including solid and liquid NMR techniques, fluorescence, infrared, and a host of other techniques. A recent review article covers most of the past and recent results of spectroscopic investigations[7] amongst other topics concerning the fundamentals of bonded phase chromatography.

Although a number of these techniques are quite capable of yielding specific analytical information concerning bonding information, for example infrared on the nature of hydroxyl groups[8] and solid NMR on the hydroxyl environment of surface silicons[9-11], the signals from these experimental techniques includes all conformations at once within the experimental measurement time. This tends to make lines broad because each spectroscopic peak represents the distribution function of all of the conformations present. There appears to be no current experimental methodology which can resolve all of the individual conformations and give detailed information on the dynamics and conformations of the bonded phase; however, NMR experiments are clearly capable of distinguishing between motions that are fast and those that are slow.

In this paper we will focus on the conformation of the C18 phase. Our approach is the reverse of laboratory experiments; instead of observing all the various conformations at once and then trying to unravel the information, we will generate the different conformations using modern computational chemistry simulation techniques. In this manner, the complexity of the many individual chain conformations, which are present at one time, is easily studied by allowing an atomistic–based computer simulation technique to generate the molecular motion. The solvent will be explicitly simulated in this study; various fractional compositions of methanol-water mixtures are used to examine the bonded phase conformation.

Computational Methodology. The technique used to simulate the atomic motions of bound alkyl chains is called Molecular Dynamics (MD). A complete overview of this technique may be found in reference 12. In MD, one describes the system to be simulated in a unit cell filled with atoms which are described by Cartesian coordinates, a connection table (giving atom–atom bonding connections), atom types (carbon, nitrogen, etc.) and bond orders (single bonds, double bonds, etc.). This unit cell is then subjected to a thermal bath and the atoms allowed to move according to the classical (Newton's) laws of motion. The potential energy of this system is described by a sum of energy terms for bond length, bond angle, dihedral (torsional) angle, electrostatic, and non-bonded van der Waals potentials. These potentials are obtained from experiment and ab initio calculations, and are discussed in the context of molecular mechanics[13]. For the simulations reported here we have used AMBER[14] potentials with the TIP3[15] water model The calculations include energy minimization prior to MD utilizing some fixed atoms in the silica substructure and periodic boundary conditions (PBC's) described below. The main body of code used for MD is the Batchmin[16] package, modified in our laboratory to include PBC's.

Preparation of the Unit Cell. Modeling this system requires model stationary phase structures. Towards this end we produce silica fragments which contain a variety of Si-O ring sizes from 4 to 8 rings and according to a distribution function obtained from simulation[17]. The silica surface layer structure is produced by one of many FORTRAN-77 programs which were written for this project. Upon ring addition, performed via a Monte Carlo technique, the bonded alkyl chain(s) are added by removing a hydrogen from a silanol group and preserving the O-Si-O-H dihedral angle. This structure is then energy minimized using the conjugate gradient method[18] prior to dynamics. The silicon atoms in the silica lattice are held stationary to mimic the hindered movement of a multilayered surface by applying a deep and narrow Hookean well potential; note that the oxygen atoms in the silica are not held fixed.

It is recognized[19] that adsorption phenomena on surfaces with defects (as is the case of the amorphous silica model used here) presents a much different surface to adsorbed or bound chains than a periodic, crystalline surface; this is one of the reasons for using an amorphous surface model rather than using a repeating crystalline surface. Furthermore, the chains are packed so that there is a higher density in the center than at the edges where chains will move sideways because their neighbors are farther

away. This is accomplished by adjusting the cell length used in the PBC's
to obtain the desired bonding density, although the rough bonding density is
determined through random chain placement in the computer program that
builds the silica and chain system. Percolation theory suggests that it is un-
realistic to assume a constant distance between chains for a random growth
process like the synthetic procedure used to bind chains to surfaces. In this
regard, we have purposely biased the chain positions to be more dense in the
middle of the cell as compared to the cell edges. Lochmüller and coworkers[20]
have discussed this point previously. No attempt has been made in these
structures to adequately include the surface curvature and morphological
effects which are present in real silica particles; transmission electron microscopy of chromatographic silica[21] show these effects through the presence
of fundamental microgel beads of the order of about 100 Å diameter. The
cell size used here is 32 Å per side. This yields an average chain loading of
2.1 μmoles per square meter.

To add the solvent to the energy minimized silica and chain structure, a lattice-blocking technique is used where overlapping van der Waals
considerations dictate whether the solvent molecule is added or the lattice
block left empty. The number of methanol and water molecules added to
the silica and chain structure is determined by the required concentration
ratio and is available in high accuracy[22]. Upon solvent addition, the system
is energy minimized prior to molecular dynamics.

Molecular Dynamics (MD). The MD simulations are performed in the
constant volume and temperature ensemble using PBC's and typically run
for 60 ps (picoseconds) of physical time and at 300 °K. Approximately the
last 30 ps are used to obtain the radial distribution functions described below.
PBC's allow a large scale system to be simulated by eliminating or minimiz-
ing edge effects; when an atom moves out of one face of the cell, it reenters
the face parallel to it. When the potential functions are set to be of range
less than half the box length, then atoms are always surrounded by some
neighbors. A good discussion of the PBC technique is had in reference 12.
The constant temperature bath is set for reequilibration at 0.2 ps intervals,
the step size for integration of the equations of motion is 1 femtosecond and
the SHAKE algorithm[23] is used to constrain all O–H and C–H bond lengths
(but not angles). Potentials are gently rolled off to prevent propagation of
spurious higher harmonics using a procedure described in reference 24.

Results and Discussion

Multiple C18 chains - no solvent. The structures for the C18 ligands
bound to silica are shown in figures 1 and 2 at 0.5 and 50 ps, respectively, in
stereo view. These figures can be viewed in three dimensions using viewers
that are commonly used for biochemical applications. Note that the hydrogens are deleted from the figures for easier viewing. Constant temperature
was achieved in about 15 ps of simulation.

The ligands in the most concentrated region of figure 2 stand up
normal from the surface. This is no surprise because it is well known in the
polymer literature on steric stabilization of colloids by bound chains that
excluded volume effects drive chains towards this position. When viewing
these results in dynamic viewing mode there appears to be a large amount
of motion in the chains, especially for the carbon atoms farthest from the
surface. Chains which are present in low density regions lie perpendicular
to the surface when not encumbered by the effects of excluded volume - this
is shown more dramatically in the single chain simulations described below.
Although the experiments conducted without solvent have little significance
in separation science, it is interesting to note that there is considerable motion in the chain system. This suggests that the chains behave more as a

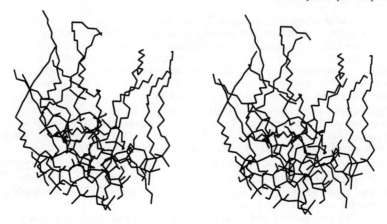

<u>Figure 1</u>. Multiple C18 chains with no solvent at 0.5 ps of molecular dynamics.

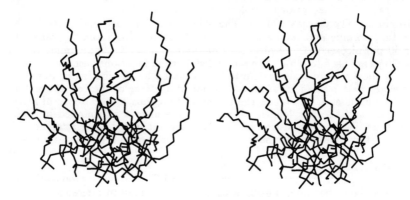

<u>Figure 2</u>. Multiple C18 chains with no solvent at 50 ps of molecular dynamics.

two-dimensional liquid near room temperature and not as a crystalline solid phase.

Single C18 chain - no solvent. As shown in figure 3 (0.5 ps) and figure 4 (50 ps) the chain rapidly self-adsorbs to the surface of the silica. Note that the hydrogens are explicitly included in these figures to show the various gauche and trans effects of the torsional angles on the chain. The excluded volume effect which drives chains towards the surface normal position is absent in these cases. Numerical experiments for the single chain C8 case do not show the self-adsorption because there isn't enough carbons for this type of flexibility. This suggests that for C18 in the dry state, regions of low bonding density will have chains that are adsorbed to the surface; regions of high density will show the excluded volume effect and the chains will stand normal to the surface.

C18 chains - 20% Water 80% Methanol. The chain conformation for this case is shown in figure 5 at 50 ps with the water and methanol removed from the figure to facilitate viewing the chain conformation. Again, the

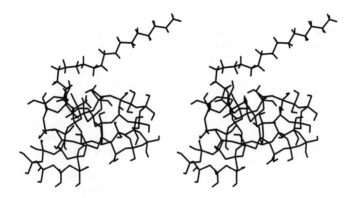

Figure 3. A single C18 chain with no solvent at 0.5 ps of molecular dynamics.

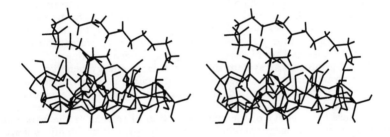

Figure 4. A single C18 chain with no solvent at 50 ps of molecular dynamics.

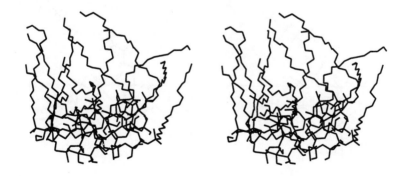

Figure 5. Multiple C18 chains with 20% Water 80% Methanol at 50 ps of molecular dynamics. The solvent has been removed for visualization.

chains are mostly normal to the surface. Dynamic viewing of the simulation shows that the the chains appear to be damped by viscous effects and appear to undergo less rotationally–driven configurations per unit time than in the dry state. Still, excluded volume drives the chains to the mostly normal position with respect to the surface in the most concentrated portion of the system.

For chain systems run in solvent it is possible to analyze where the solvent is positioned with respect to the chains. This has been done for the water–chain and methanol–chain ensembles in the form of pair radial distribution functions[12], $g(r)$. These functions give the probability density for finding the water and methanol molecules at a certain distance from the carbons on the chain. Upon analysis of $g(r)$ for multiple chains in 20% Water 80% Methanol, it is found that the water is preferentially depleted in the chain vicinity away from the silica surface showing a distinct hydrophobic effect. This indicates that while methanol is roughly at its bulk fluid density near the chains, a dense gas picture of methanol is probably more accurate in describing the chemical environment near the chains. This has profound consequences for the interpretation of the LC experiment in water–methanol solutions as will be discussed below.

C18 chain - 20% Water 80% Methanol. Figure 6 shows the chain conformation after 50 ps of simulation with the same structure used in the single chain study above and with the solvent removed for visualization. In this case the chain does not go through adsorption but rather appears to be approximately parallel to the surface. There are two reasons that can be suggested as to why the chain does not self–adsorb in this case. First, as viewed by the silicate oxygen-water $g(r)$, the surface is loaded with water which forms compact tightly-bound multi–layers through hydrogen bonding with the silica hydroxyl groups. The system must pay a large energy penalty to break this water structuring so the adsorption of the chain is unfavorable through non–bonded interactions. Second, the chain itself maintains an excluded volume with respect to solvent ordering and hence to maintain this excluded volume where the vicinity of the chain is devoid of solvent at the bulk multicomponent fluid density, the chain must keep away from highly localized pockets of water. It is well–known that very good qualitative interpretations of the hydrophobic effect can be obtained with the simple potential energy functions utilized in these force fields.

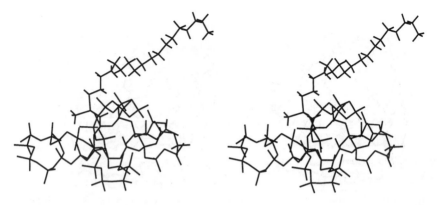

Figure 6. A single C18 chain with 20% Water 80% Methanol at 50 ps of molecular dynamics. The solvent has been removed for visualization.

Decane in Water–Methanol Mixtures. A more fundamental experiment which aids in the interpretation of the bonded chain results is given in figure 7, where $g(r)$ of decane in water–methanol is presented. In the four cases shown in figure 7, two different pairs are used to construct $g(r)$ at two different methanol–water concentrations. In all cases one of the pairs is the

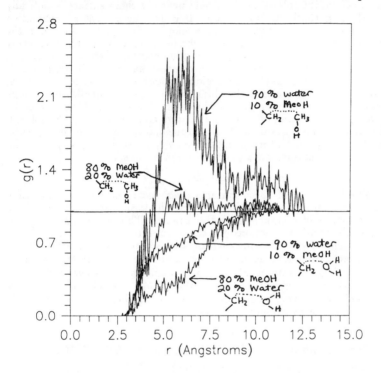

Figure 7. The pair radial distribution function, $g(r)$, of decane in water–methanol.

methyl and methylene carbons on the C18 chains. The other pair member is either the oxygen on water or the methyl carbon on methanol. In these cases $g(r)$ can be obtained at any chain carbon position, or as shown in figure 7, $g(r)$ can be obtained as a summed result over all chain carbon atoms. This experiment is much simpler to interpret than $g(r)$ in the presence of silica because the silica introduces preferential adsorption of water over methanol, as seen from $g(r)$ plots of the solvated chain system described above. In addition, the decane is insoluble in water–methanol mixtures (even in pure methanol), as is the C18 molecule.

The interpretation of figure 7 is rather straightforward. For high water concentrations (90% Water 10% Methanol), the methanol concentration is preferentially enhanced over its bulk fluid concentration (where $g(r) = 1$). In addition, under high water concentrations, the water is displaced away from the chain and only reaches the bulk density approximately 10 Å from the carbon atoms. Conversely, for high organic solvent content (20% Water 80% Methanol), only a slight enhancement of methanol is seen over the bulk density of methanol. However, in this case a very significant depletion of water takes place near the alkyl chain, again suggesting that the solvent environment near the chain is significantly different than in bulk fluid and probably resembles the dense gas mentioned previously.

These results suggest that when a solute molecule is diffusing into the chain system, the solvent environment near the chain surface is radically different than in the bulk fluid phase. Because the solvent concentration gradient extends between two and three coordination shells away from the alkyl chain, these results suggest that the chain–fluid interfacial fluid structure may substantially contribute to the specific retention behavior of solutes in reversed–phase HPLC. For instance, solutes that are very hydrophobic may be driven into the alkyl chain region through the action of the solvent structure at the chain while hydrophilic solutes may never diffuse through the solvent gradient and interact with the chain. The extent of this interfacial driving force may be moderated by the relative concentration of the solvent, as is the case during a gradient HPLC experiment. This point is currently being tested in our laboratory using free energy perturbation and potential of mean-force calculations on a select group of solutes.

Conclusions

The full story of these systems does not come from visualization alone, but rather from a description of where the local solvent concentration lies as a function of distance from the chains. This is expressed by the pair radial distribution functions for the species of interest. Nonetheless, visualization shows gross features which are easy to understand and these include: 1) excluded volume is most important in the conformational analysis of concentrated chain systems, 2) For dilute chain systems, solvent effects are most important in determining chain conformation. Dry chains can adsorb yet methanol-water mixtures do solvate the chains, albeit most of this is still through balances of hydrophobic effects.

REFERENCES

1. D. E. Martire and R. E. Boehm, J. Phys. Chem., 1983, 87, 1045.

2. K. A. Dill, J. Phys. Chem., 1987, 91, 1980.

3. C. Horvath, W. Melander and I. Molnar, J. Chromatogr., 1976, 125, 129.

4. J. G. Dorsey and K. A. Dill, Chem. Rev., 1989, 89, 331.

5. K. B. Lipkowitz, J. M. Landwer, and T. Darden, Anal. Chem., 1986, 58, 1611.

6. K. B. Lipkowitz, D. A. Demeter, R. Zegarra, R. Larter, and T. Darden, J. Am. Chem. Soc., 1988, 110, 3446.

7. L. C. Sander and S. A. Wise, CRC Crit. Rev. in Anal. Chem., 1987, 18, 299.

8. J. B. Peri and A. L. Hensley, J. Phys. Chem., 1968, 72, 2926.

9. D. W. Sindorf and G. E. Maciel, J. Am. Chem. Soc., 1983, 105, 3767.

10. D. W. Sindorf and G. E. Maciel, J. Am. Chem. Soc., 1981, 103, 4263.

11. D. W. Sindorf and G. E. Maciel, J. Am. Chem. Soc., 1983, 105, 1487.

12. M. P. Allen and D. J. Tildesley, 'Computer Simulation of Liquids', Oxford, New York, 1987.

13. U. Buckert and N. L. Allinger, N. L., 'Molecular Mechanics', ACS Monograph 177, American Chemical Society, Washington, D.C. 1982.

14. S. J. Weiner, P. A. Kollman, D. A. Case, U. C. Singh, C. Ghio, G. Alagona, S. Pofeta, and P. Weiner, J. Am. Chem. Soc., 1984, 106, 765.

15. W. L. Jorgenson J. Chem. Phys., 1982, 77, 4156.

16. F. Mohamadi, N. G. J. Richards, W. C. Guida, R. Liskamp, M. Lipton, C. Caufield, G. Chang, T. Hendrickson, and W. C. Still, J. Comp. Chem., 1990, 11, 440.

17. B. P. Feuston and S. H. Garafolini, J. Chem. Phys., 1988, 89, 5818.

18. W. H. Press, S. A. Teukolsky, W. T. Vetterling, and B. P. Flannery, 'Numerical Recipes', 2nd edition, Cambridge University Press, New York, 1992.

19. M. R. Wattenbbarger, V. A. Bloomfield, and D. F. Evans, Macromolecules, 1992, 25, 261.

20. C. H. Lochmüller and D. R. Wilder, J. Chromatogr. Sci., 1979, 17, 574.

21. N. Tanaka, K. Kimata, T. Araki, H. Tsuchiya, and K. Hashizume, J. Chromatogr., 1991, 544, 319.

22. G. C. Benson and O. Kiyohara, J. Solution Chem., 1980, 9, 791.

23. J.-P. Ryckaert, G. Ciccotti, and H. J. C. Berendsen, J. Comp. Phys., 1977, 23, 327.

24. M. Prevost, D. Van Belle, G. Lippens, and S. Wodak, Mol. Phys. 1990, 71, 587.

^{2}H and ^{13}C NMR Studies of Reversed Phase Liquid Chromatographic Stationary Phases: Solvation and Temperature Effects

K. B. Sentell,* D. M. Bliesner,[1] and S. T. Shearer

DEPARTMENT OF CHEMISTRY, UNIVERSITY OF VERMONT,
BURLINGTON, VT 05405-0125, USA

[1]Present Address: Zeneca Pharmaceuticals Group, Zeneca Inc., NLW2, Wilmington, DE 19897

1 INTRODUCTION

Stationary phase conformation and solvation have been shown to play important roles in chromatographic behavior in reversed phase liquid chromatography (RPLC);[1-8] as a result these are very active areas of fundamental research. Studies of RPLC bonded phase structure and solvation are complicated because there is no true "macroscopic phase"; i.e. its composition and properties vary according to microscopic location on the silica surface as well as with alkyl ligand depth. Another complexity for thermal studies on these materials is that they are "solids" which are not in thermodynamic equilibrium with respect to phase changes (similar to glassy polymers[9]). Stationary phase structure and solvation have primarily been studied via chromatographic experiments. However, the disposition of such measurements dictates that information about the nature of the stationary phase be drawn by inference, via interpretation of the chromatographic results. In contrast, spectroscopic techniques afford a much more direct means of studying the stationary phase surface. Just as importantly, spectroscopic measurements can be made independently of chromatographic measurements; their orthogonality often leads to information that is complementary to that obtained chromatographically. The combined approach provides a much more complete and realistic picture of molecular level contributions to RPLC retention and selectivity than can be obtained from either technique alone[10].

Solution state ^{13}C NMR measurements on these materials provide information about motions in the grafted alkyl stationary phase ligands that are in the range of the Larmor frequency of the ^{13}C nuclei (MHz). These experiments are essential for determining bonded phase mobility under solvated conditions, which much more nearly approximate actual RPLC systems. Although not as many individual resonances of distinct bonded ligand sites can be delineated as in solid state NMR experiments, useful information about the dynamics of at least three different regions in a typical bonded phase alkyl ligand can be obtained from solution state experiments. Additionally, solution state ^{2}H NMR measurements can be carried out on perdeuterated mobile phase components in order to study their ability to solvate RPLC bonded phases. The most useful parameters measured in either type of experiment are the spin-lattice relaxation times, T_1, of the nuclei of interest, since they are inversely proportional to τ_c, the nuclear correlation time. The correlation time is a measure of how long it takes the nucleus to rotate through one radian and is therefore a measure of molecular motion. In the solution state, as the correlation time increases (e.g. molecular motion decreases), the longitudinal relaxation time, T_1, decreases.[11] Greater molecular mobility is therefore indicated by an increase in T_1. We have used these types of experiments in our laboratory to study the effects of mobile phase composition and temperature on the solvation and mobility of C_{18} RPLC bonded phases. By carrying out these studies under conditions which closely approximate actual RPLC operating conditions, our goal was to use NMR measurements to better understand the chromatographic process at the molecular level.

2 EXPERIMENTAL

NMR Measurements

T_1 values for deuterium oxide (D_2O), methanol-d_4 and acetonitrile-d_3 in methanol/water and acetonitrile/water mobile phase mixtures, both for the binary solutions alone and in contact with stationary phase, were measured with a Brüker WM-250 NMR spectrometer operating at a field strength of 5.875 Tesla. The standard inversion recovery sequence [(180° - τ - 90° - Acquire - Delay)$_n$] was used to acquire the T_1 values, where τ ranged from 0.005 to 2 s for D_2O and from 0.025 to 24 s for the organic solvents. The sequence employed no less than 10 τ values and each acquisition necessitated from 2 to 32 scans. A relaxation delay in excess of 5 T_1 was used to allow the system to return to equilibrium. Total acquisition times ranged from 10 to 50 minutes. Care was taken to minimize reflected power for each sample in order to minimize errors in the 180° and 90° pulses. All spectra were acquired using automatic field frequency lock, and temperature was held constant at 30 °C (303 K).

^{13}C T_1 values for carbon nuclei at various positions along the alkyl ligand of the C_{18} RPLC bonded phase LT1 (described below) were measured using a Varian 300XL spectrometer operating at a field strength of 7.02 T. The analogous ^{13}C T_1 values for the neat dimethyloctadecylchlorosilane were measured on the Brüker WM-250 spectrometer. The fast inversion recovery sequence[12] was used to acquire T_1 values, using τ values ranging from 10 μs to 32 s. The sequence employed no less than 8 τ values and each acquisition necessitated a minimum of 16 scans. All spectra were obtained under conditions of broad band decoupling. The temperature was varied from 273 to 365 K in 3-5 K increments, and was measured by placing a cromel thermocouple in a sample of the same composition as the experimental sample and lowering it into the bore of the magnet. Temperatures are reported to ± 3 K; more accurate measurements are not possible due to temperature variations within the sample caused by decoupling and the application of the pulse sequence.   Final   sample   heights   for   all measurements were ≈ 8 cm. This allowed irradiation of the entire sample by the transmitter coil, greatly enhancing signal to noise. All T_1 values were calculated by manual measurement of peak height for each τ value, followed by a three-parameter least squares exponential fit of the peak height versus τ plot. Relative error for replicate measurements was in all cases less than 5%. All T_1 values are stated at a confidence level of plus or minus one standard deviation.

Chemicals

99.98% deuterium oxide (Aldrich, Milwaukee, WI), methanol-d_4, acetonitrile-d_3 (Isotec, Miamisburg, OH), dimethyloctadecylchlorosilane (DOCS; Hüls America, Bristol, PA) and HPLC grade methanol and acetonitrile (Fisher Scientific, Fair Lawn, NJ) were used without further purification. HPLC grade water was obtained in-house using a Nanopure (Sybron Corp., Boston, MA) water purification system. Chromatographic solutes were obtained from Sigma Chemical Co. (St. Louis, MO) and Aldrich Chemical Co. Two silica-based C_{18} stationary phases were also used in these studies. The first, designated LT1, is a 10 μm monomeric non-endcapped phase with very high octadecyl ligand bonding density (4.4 μmol/m^2). LT1 was synthesized in our laboratory using DOCS under conditions which have been previously described.[8,13] The second stationary phase, designated ODS-1, is a commercially available (Spherisorb S5-ODS-1; Phase Separations, Norwalk, CT), 5 μm non-endcapped phase with a much lower C_{18} bonding density (1.41 μmol/m^2). For ODS-1, the octadecyl ligand is attached to the silica surface using a polyfunctional C_{18} silane under anhydrous conditions.

Sample Preparation

^{2}H Studies. Approximately 3% (by volume) D_2O, CD_3OD or CD_3CN was added to its nonlabelled analog in a chromatographic mobile phase reservoir. All mobile phases were degassed by sonication under vacuum. The desired mobile phase mixture (volume-to-volume percent ratio of organic solvent to water) was delivered into a 10 mm NMR tube by two Waters (Waters Assoc., Milford, MA) Model 510 chromatographic pumps controlled by a Waters Model 680 automated gradient controller. For both the methanol:water and acetonitrile:water mixtures, samples were prepared in 10% or smaller volume/volume (v/v) organic/water increments ranging from 100:0 v/v% to 0:100 v/v%. ^{2}H T_1 for the deuterated component was then measured for each binary mobile phase mixture.

To insure proper wetting of the stationary phase with the labelled mobile phase solutions in the second half of the experiments, mobile/stationary phase mixtures were prepared in the following manner. First, the chromatographic stationary phase was dried under vacuum in excess of 12 hours at 110 °C to ensure that all physisorbed solvents were removed. A 1.1 g portion of the dry stationary phase was hand-packed into an empty 10 cm X 4.6 mm stainless steel chromatographic column. Mobile phase of the desired volume percent ratio was then pumped through the column at a flow rate which would generate at least 1500 psi of backpressure. Approximately 50 ml of solvent was pumped through the column in one direction, the column was reversed and an additional 50 ml was pumped in the opposite direction. Following the wetting procedure, one of the column end fittings was removed from the column and the wetted stationary phase was pumped as a smooth paste into an 8 mm NMR tube. The T_1 experiment on the paste was then performed as described above for the solvent samples. It should be noted that at mobile phase compositions containing a very low volume percent of organic solvent, considerably larger amounts of mobile phase (in excess of 800 ml) were needed to successfully wet the stationary phases.

^{13}C Studies. The techniques used to make the LT1 stationary phase slurry samples were those established by Bayer and co-workers, with the following modifications.[14] LT1 was added to a 10 mm NMR tube, followed by a volume of 80:20 methanol-d_4:H_2O sufficient for the liquid to cover the top of the stationary phase. The sample was sonicated in an ultrasonic bath for 30 minutes to insure wetting and allowed to sit for several hours. Next, argon gas which had been dried and de-oxygenated was bubbled through the stationary phase slurry to remove any residual dissolved oxygen.

For the neat silane measurements, the solid silane was placed in a 10 mm NMR tube under a dry argon atmosphere. The tube was capped and placed in warm water (315 K) in order to melt the silane, which was then degassed in the manner described above.

Chromatographic Measurements

An HPLC pump, injection valve, gradient controller and keypad, variable wavelength UV-visible detector operated at 254 nm (Scientific Systems Inc., State College, PA) and a chart recorder (Fisher Scientific, Fair Lawn, NJ) were used to obtain all liquid chromatographic measurements. The columns utilized in these studies (4.6 mm X 20 cm and 4.6 mm X 15 cm respectively for the ODS-1 and LT1 phases) contained the same two stationary phases as utilized in the NMR experiments. Columns were equilibrated with at least 125 ml of the mobile phase and were thermally equilibrated for at least an hour prior to chromatographic measurements. Temperature control was accomplished by passing a thermostatted solution of methanol and water through a well-insulated water jacket surrounding the column. Actual temperature was verified with a thermocouple placed within the water jacket in direct contact with the chromatographic column, with fluctuations never varying more than ± 0.2 °C. Retention data were obtained at temperatures ranging from -20 to 70 °C under isocratic conditions with 80:20 methanol:water as the mobile phase. Capacity factors were measured from triplicate injections of methanol solutions of each solute and are reported as the average of the three measurements; precision for replicate measurements was in no case worse than ± 4.0% RSD. Selectivity factors, α, were calculated from the ratio of the average capacity factors of the solute pair.

3 ^{2}H SOLVATION STUDIES

To study the degree of interaction between mobile phase components and the stationary phase, solution state NMR T_1 studies of nuclei present in the mobile phase can be applied. Although NMR measurements made on mobile phase components cannot generally be used to <u>quantitate</u> the <u>amount</u> of sorbed organic modifier in the stationary phase systems, they are valuable in a <u>qualitative</u> sense for determining the <u>extent</u> to which mobile phase components interact with the stationary phase. For any quadrupolar nucleus, the spin-lattice relaxation time (T_1) is dominated by intermolecular effects; other relaxation mechanisms such as intramolecular dipole-dipole interactions and quenching by paramagnetic impurities compete poorly if at all.[15] As the solution state correlation time for a quadrupolar nucleus increases (e.g. molecular motion decreases), T_1 decreases. Therefore T_1 values for quadrupolar nuclei can serve as sensitive measures for monitoring the binding of a molecule of interest to other species, even if the amount bound is a relatively small fraction of the total species.[15]

Deuterium (^{2}H) is a quadrupolar nucleus which has enjoyed wide use in NMR studies of intermolecular interactions in chemical and biological systems. We have measured ^{2}H T_1 for methanol-d_4, acetonitrile-d_3 and D_2O in binary aqueous mixtures and for those mixtures in contact with monomeric C_{18} bonded phases to study mobile phase/stationary phase interactions in RPLC.[10,16,17] The sample preparation method described in the experimental section equilibrates the stationary and mobile phases at pressures comparable to chromatographic operating conditions; therefore our experiments mirror wetting in real chromatographic systems, particularly for highly aqueous mobile phase mixtures.

We first measured ^{2}H T_1 at 10% or smaller volume increments over the entire binary composition range for the methyl deuterons of methanol-d_4 and acetonitrile-d_3 and for D_2O in the binary solvent mixtures. By next combining stationary phase with the binary solutions at pressures comparable to those under chromatographic operating conditions, and re-measuring the longitudinal relaxation times, the <u>change</u> in the ^{2}H T_1 for the solvent component in contact with the stationary phase versus that observed in the solution alone can be measured for both components in the binary solutions over the entire composition range. This serves to normalize the relaxation data for the former case, because bulk solution phenomena that will affect the observed T_1 in both systems are accounted for by the bulk solution measurements.[17]

The assumptions that the exchange of mobile phase species between those in bulk solution and those associated with a solid surface are rapid,[18] and that the resident lifetime of any associated mobile phase species will be much shorter than its corresponding T_1[15,18] have been shown to be appropriate for solvent interactions with microporous silica gels[15,19] as well as similar solid materials.[18,20] In these two-site rapid-exchange systems, the observed relaxation <u>rate</u> is:[15,18,20]

$$\frac{1}{T_{1(obs)}} = (F_b)\frac{1}{T_{1b}} + (1 - F_b)\frac{1}{T_{1s}} \qquad (1)$$

where $T_{1(obs)}$ is the observed T_1 value of the mobile phase component when the solution is placed in contact with the stationary phase material, F_b is the fraction of mobile phase molecules in the associated state, T_{1b} is the relaxation time for the associated species and T_{1s} is the T_1 value in bulk solution. When the fraction of molecules in the associated state that determines the observed T_1 is small, and the relaxation rate of the associated D_2O molecules is greater than for those in the bulk solution, eq. 1 is valid.[15,20] The observed T_1 values are a weighted average of those for free solution species and for those species associated with the stationary phase.[15] Since deuterated components make up at most 3% of the total volume of the bulk mobile phase, they will provide an even smaller fraction of the species associated with the stationary phase. Moreover, the ^2H relaxation times measured for the solutions in contact with the stationary phases are in all cases smaller than those in the bulk solution. Therefore, the criteria for the validity of eq. 1[20] are met by our experimental system.[17]

Because of rapid exchange between species in the bulk solution and the stationary phase, T_{1b} cannot be directly measured, and rigorous determination of F_b is prevented by the experimental uncertainties inherent in attempting to accurately measure small amounts of adsorbed mobile phase species on RPLC stationary phases.[21] Therefore reliable quantitative information about the amount of D_2O associated with the stationary phase cannot be obtained from our T_1 measurements. However, via eq. 1, the difference between the $T_{1(obs)}$ and T_{1s} values at each mobile phase composition permits qualitative trends in the relative degree of association of mobile phase components with the stationary phase to be gauged as a function of bulk mobile phase composition.[17] Such an association will be accompanied by reduction in diffusional freedom in moving from the three-dimensional bulk fluid to a two-dimensional surface.[15] Reduction in motional freedom of the associated mobile phase species relative to those in the bulk solution will be manifested by a decrease in the T_1 for the former species compared to that for the latter.[15] Therefore qualitative information about the relative fraction of a solution species associated with the stationary phase at that solution composition is provided by the magnitude of the decrease in the component's T_1 when the solution is placed in contact with stationary phase.[17] Similarly, when all experimental variables except for the surface morphology of the stationary phase are held constant, comparison of solution species relaxation times for identical solutions in contact with different stationary phases is possible. A greater change in the T_1 for one stationary phase system compared to that for another implies that there is a larger degree of mobile phase species association with the former stationary phase.[10]

Our measurements of the change in the ^2H T_1 (i.e. T_1 for the solution minus T_1 for the solution in contact with the bonded phase; hereafter referred to as ΔT_1) for each of the individual solution components as a function of binary solution composition have confirmed that organic modifier interactions with monomeric C_{18} stationary phases are much more complex than aqueous interactions.[16,17] Additionally, they are distinctly different for methanol when compared to acetonitrile. For both the high (4.4 $\mu mol/m^2$; LT1) and low (1.41 $\mu mol/m^2$; ODS-1) chain density C_{18} phases in contact with methanol/water mobile phases, the degree of water associated with the stationary phase was limited and almost constant. However, the degree of methanol association was considerably greater and exhibited dependence on the amount of methanol present in the bulk mobile phase. The methanol-d_4 ΔT_1 for bulk mobile phase compositions ranging from $\approx$ 30% to 80% methanol was relatively constant and was less than ΔT_1 for any other range of bulk methanol-water compositions for both stationary phases. It has been proposed by Katz *et al.* that in this composition region, water-methanol associated complexes predominate in methanol-water binary solutions.[22,23] Hydrogen bonding interactions are stronger in water-methanol associated species than in self-associated methanol species. Solution interactions for the former are also stronger and more directed than the predominantly dispersive interactions which could occur between methanol and C_{18} stationary phase ligands.[16] Additionally, in this composition range, as indicated by the extremely small ΔT_1 values measured for D_2O, accessibility to the support surface silanols is minimal.[17] In light of these considerations, the limited and relatively constant degree of methanol association with the stationary phase expected and observed over this mobile phase composition range is reasonable.

For mobile phase compositions with less than 30% methanol, a marked increase in methanol-d_4 ΔT_1 and a smaller increase in the D_2O ΔT_1 are observed. Contact angle measurements have confirmed that pure water does not wet C_{18} bonded silica surfaces and that the addition of 20% methanol to pure water only brings about partial wetting.[24] The hydrophobic alkyl stationary phase chains are thought to assume a highly collapsed or folded configuration in predominantly aqueous mobile phases, in order to minimize their surface contact with the polar mobile phase.[4,5,24] Mobile phase components become entrapped within the collapsed stationary phase structure (most likely within narrow-necked silica pores) and this results in a distinct change in their T_1.[10,16,17]

When the bulk mobile phase contained greater than 80% methanol, the methanol ^2H ΔT_1 increased with volume percent of methanol, and its magnitude implied a significantly greater degree of methanol association with the stationary phase than exhibited in the intermediate composition region.[16] The volume fraction of the less polar self-associated methanol species in bulk solution increases linearly at nominal methanol volume fractions above $\approx$ 0.8, commensurate with the decrease in water-methanol associated complexes.[22,23] The self-associated methanol solution species present in this composition region are much more likely to experience dispersive interactions with the C_{18} chains of the stationary phase,[25-27] since most of the methanol is not hydrogen bonded with water. Solution state ^{13}C NMR spectra acquired in our laboratory for the LT1 phase in contact with neat methanol exhibit an additional resonance at a chemical shift corresponding to that for the silyl methyl groups.[28] This indicates that under these conditions the locations very near the attached end of the alkyl ligand are exhibiting more liquid-like behavior. This is suggestive of decreasing steric hindrance to motion near the silica surface, which might be attributed to alkyl chain extension.[10] However, the D_2O ΔT_1 behavior indicates that the degree of D_2O association with the stationary phase is little changed over this composition region.[17]

Consequently, any increase in stationary phase chain extension from methanol uptake is not so pronounced as to allow surface silanol groups to effectively compete for hydrogen bonding interactions with solution species. Therefore the much larger increase in methanol-d_4 ΔT_1 in this range is more logically attributed to more effective dispersive interactions between the bonded phase alkyl ligands and the self-associated methanol species.[17] In total, these experiments indicate that when methanol-d_4 and D_2O are placed in contact with monomeric C_{18} stationary phases, changes in T_1 for both of these species are largely due to bulk solution microstructure and its resultant effects on bonded phase solvation and structure.[10]

For acetonitrile/water contact solutions, the degree of water association with the stationary phase was a function of bulk mobile phase composition, whereas the degree of association of acetonitrile was relatively constant regardless of the mobile phase composition. This was opposite from the case with methanol/water mobile phases[17]. When added to bulk aqueous solutions, acetonitrile enters cavities in the well-defined water structure until these sites are occupied. Acetonitrile becomes increasingly self-associated in aggregates or loosely defined clusters as its amount in the bulk solution is further increased.[29,30] Although this results in bulk solution microheterogeneity, the acetonitrile species, because of their extensive self-association, experience a relatively homogeneous solution environment over a large binary composition range.[22,30,31] Acetonitrile-d_3 therefore displays a relatively constant degree of association behavior (ΔT_1) with C_{18} stationary phases throughout the composition range.[16]

For both stationary phases, the degree of association was much less for water than for acetonitrile. However, ΔT_1 behavior for D_2O in contact with both stationary phases changed rapidly in the solution composition region from 80-100% acetonitrile. This cannot be due to solution properties in the absence of the stationary phase surface, since the physical properties of acetonitrile-water mixtures change slowly and monotonically over the entire composition range.[15] The ^2H chemical shifts observed in our T_1 studies for D_2O in contact with the bonded phase materials were in all cases virtually identical to the corresponding chemical shifts in the mobile phase binary solutions.[17] This indicates that extreme changes in the chemical environment of the D_2O molecules, such as would be expected if the majority of them were directly hydrogen bonded to surface silanols, are not present.[15] Increased association of water with the C_{18} bonded phases in this composition region is therefore the result of a combination of hydrogen bonding of D_2O with residual silanols and the formation of water-acetonitrile mixed species complexes in the acetonitrile-rich alkyl chain environment.[15] At highly aqueous compositions, acetonitrile is likely entrapped within narrow-necked silica pores by a relatively collapsed chain structure,[5] just as is the case for comparable methanol compositions. The D_2O results that we have obtained[17] for mobile phases containing in excess of 50% by volume of acetonitrile are in excellent agreement with the findings of Marshall and McKenna from similar NMR studies.[15]

It is also instructive to compare the association of the organic modifiers versus the aqueous component for our NMR studies of the mobile phases in contact with the stationary phases.[10,16,17] A larger extent of association was observed for acetonitrile than for methanol for both stationary phases.[16] There was more difference between the low and high ligand density stationary phases in the degree of association for the organic modifier mobile phase components (especially acetonitrile) than for the aqueous component.[10] The effects of chain density on the formation of the solvation layer in RPLC therefore appear to be more important for less polar mobile phase components such as methanol or acetonitrile than for water. This further supports earlier chromatographic observations on the partitioning behavior of organic modifier components of the mobile phase within RPLC stationary phases.[32] Our ^{2}H NMR work reinforces the view that the association of mobile phase components with the stationary phase that causes the formation of a solvation layer in RPLC is a complex phenomenon that is largely determined by the competition between the relative strengths of solvent/solvent and solvent/stationary phase interactions.[15] This is illustrated by the distinct correlations between bulk solution structure and stationary phase solvation by individual solution components.[16,17]

4 ^{13}C TEMPERATURE STUDIES

Mobile phase solvation is a crucial influence on the structure and conformation of RPLC bonded phase ligands; therefore in order to fully utilize ^{13}C NMR to better understand the effects of temperature on RPLC retention and selectivity, temperature dependent solution state NMR measurements are essential. Since in the solution state, decreasing ^{13}C T_1 values imply decreasing molecular motion, overall bonded phase ligand mobility can be examined by T_1 measurements. More importantly, specific information about the dominant modes of molecular motion for various spatial regions in the alkyl ligand can be determined, since the temperature dependence of T_1 for spin-1/2 nuclei such as ^{13}C is indicative of the primary mechanism for nuclear relaxation. Dipole-dipole and spin-rotation interactions are the predominant relaxation mechanisms for ^{13}C, with the former greatly prevailing in most cases.[11] Typical liquid state organic molecules relax via the dipolar mechanism; for this mechanism a plot of T_1 versus inverse temperature (1/T) results in a linear relationship with a negative slope.[33] Spin-rotation is for the most part a secondary relaxation mechanism for spin-1/2 nuclei. However, for small top symmetrical molecules or rapidly rotating functional groups, spin-rotation can contribute to overall relaxation to an appreciable extent.[33-35] The latter consideration is relevant to RPLC bonded phases, since previous NMR studies have indicated that at ambient temperatures or above, the motion of the terminal methyl group of RPLC alkyl ligands is dominated by rotation about the end bond.[14,36-41] When spin-rotation dominates the relaxation mechanism, a plot of T_1 versus 1/T yields a positive slope. Gillen *et al.*[33] have shown that for methyl iodide, dipole-dipole interactions dominate the relaxation of the methyl group at reduced temperatures, and this results in the negative slope typically observed in a T_1 versus inverse temperature plot for organic molecules. However, for the rapidly rotating methyl carbon the contribution of spin-rotation to relaxation is larger at higher temperatures, and this produces a slightly positive slope in this portion of the observed T_1 vs. $1/T$ plot.[33]

The above example illustrates that although it is normally regarded as a trivial contribution to ^{13}C nuclear relaxation for typical liquid state organic molecules, in some systems the contribution from spin-rotation becomes a significant part of the overall relaxation mechanism with increasing temperature, and at sufficiently high temperatures spin-rotation can dominate the relaxation process.[33] This is a relevant consideration in a chemical system such as an anchored RPLC alkyl ligand, wherein there is a possibility of contribution to relaxation from both of these mechanisms for the chain segments at the free end of the ligand. Consequently, monitoring the temperature dependence of the slope of the [13]C T_1 vs. $1/T$ plot for various carbons along the bonded alkyl ligand in the stationary phase should make it possible to observe abrupt changes in the rotational freedom of the end methyl groups as a function of temperature. Such changes would be expected if a distinct re-ordering in stationary phase structure (e.g. a second order phase transition) were taking place.[10]

Sentell and Henderson had consistently noted distinct curvature between 25 and 30 °C in van't Hoff plots of polycyclic aromatic hydrocarbon and polyarene isomers for the high alkyl chain density (4.4 $\mu mol/m^2$) C_{18} stationary phase LT1 in contact with an 80:20 methanol:water mobile phase, as illustrated for pyrene in Figure 1a.[8] Figure 1b shows that chromatographic selectivity for linear and/or planar vs. bent and/or nonplanar solute isomers also increased dramatically in this stationary/mobile phase system at temperatures below 30 °C.[8] A possible explanation for these observations was a second order phase transition of the bonded octadecyl chains from a relatively isotropic state with a large population of gauche conformers at higher temperatures to a more ordered and extended liquid-crystalline-like state with a larger population of trans conformers at lower temperatures.

In order to use a complimentary technique which would provide information on the motional effects of temperature for different segmental regions of the alkyl ligand, we have made [13]C T_1 solution state measurements on this stationary/mobile phase system over the same temperature range as

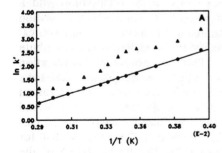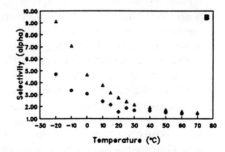

Figure 1. (a) Van't Hoff plots for pyrene with a high chain density [4.4 $\mu mol/m^2$] (▲) and a low chain density [1.5 $\mu mol/m^2$] (●) monomeric C_{18} stationary phase and 80:20 methanol:water mobile phase. Note the curvature in the plot for the high density phase. (b) Selectivity plots for p-terphenyl vs. o-terphenyl for a high [4.4 $\mu mol/m^2$] (▲) and a low chain density [1.5 $\mu mol/m^2$] (●) monomeric C_{18} stationary phase and 80:20 methanol:water mobile phase. [Reprinted from ref. 8, copyright 1991 Elsevier Science Publishers B.V., with permission].

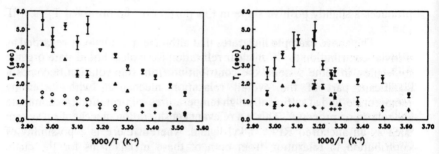

Figure 2. Plot of ^{13}C spin-lattice relaxation time (T_1) versus inverse temperature for ^{13}C resonances corresponding to alkyl ligand chain positions in (a; left) neat dimethyloctadecylchlorosilane and (b; right) a high chain density [4.4 μmol/m^2] monomeric C_{18} stationary phase (LT1) in contact with 80:20 methanol:water mobile phase. The chain segment positions plotted are C_4-C_{15} (+), C_{16} (O), C_{17} (▲) and C_{18} (▼). [Reprinted from ref. 10, copyright 1993 Elsevier Science Publishers B.V., with permission].

the chromatographic experiments.[28,42] For comparison to the bonded phase results, T_1 experiments first were performed from 5-70 °C on neat dimethyloctadecylchlorosilane (DOCS). For the neat silane, the T_1 versus inverse temperature plot resulted a negative slope for all four of the carbon resonances present in the spectrum (C_{4-15}, C_{16}, C_{17}, and C_{18}, where numbering begins at the silyl end of the alkyl chain; Figure 2a), even though DOCS is a waxy solid at temperatues ≤ 27 °C and a liquid at temperatures above.[10,42] However, Figure 2b illustrates that the ^{13}C T_1 measurements over the same temperature range for the high chain density LT1 phase slurried in 80:20 methanol: water displays evidence of a distinct change in ordering in this material. The plot of T_1 versus inverse temperature for the bulk chain (C_4-C_{15}) carbons results in a small but negative slope, in agreement with that expected for a dipolar relaxation mechanism.[28,42] In contrast, T_1 vs. 1/T plots for both the terminal methylene (C_{17}) and methyl (C_{18}) carbons demonstrate distinct curvature and an accompanying change in slope at ≈ 35 °C. This implies that although spin-rotation is the predominant relaxation mechanism for the two terminal chain segments at temperatures greater than 35 °C, dipolar relaxation is dominant at lower temperatures. This supports the proposition that the spin-rotational component of relaxation for the terminal methylene and methyl carbons is hindered at temperatures below 35 °C because of a tangible increase in ordering of the bonded phase ligands which restricts rotational mobility.[10,28,42] This is extremely relevant to chromatographic performance, since the temperature range within which the change in bonded phase chain ordering and its concomitant effects on chromatographic selectivity for isomeric compounds occurs encompasses the ambient operational temperature range in most laboratories. Our temperature dependent ^{13}C T_1 data are a further demonstration of the excellent utility of NMR measurements for providing corroborating information about stationary phase structure that is both independent from and complimentary to chromatographic measurements.

5 CONCLUSIONS

Some general conclusions about RPLC bonded phase solvation and structure can be drawn from the NMR studies described in this paper. Although solution state NMR measurements are generally inappropriate for precisely quantitating the composition of the bonded phase solvation layer, they are extremely useful for elucidating the solvation behavior of individual mobile phase components over the entire range of binary solution compositions. Our ^{2}H NMR studies have shown that the solvation environment at the bonded phase surface for any particular mobile phase composition is extremely dependent on the microstructure of the bulk mobile phase solution,[10,16,17] because association of mobile phase components with the stationary phase is largely determined by the relative strengths of solvent/solvent and solvent/stationary phase interactions.[15] Also, even though the alkyl ligands of RPLC bonded phases exhibit mobility that is in the liquid-like frequency of motion over a wide temperature range, temperature can change the degree of contribution of the rotational component of motion for the terminal methylene and methyl chain segments. Rotation about the end bonds is at least somewhat hindered at subambient temperatures, particularly for phases with high ligand surface densities, where the spatial proximity of the ligands promotes a high degree of cooperativity between neighboring ligands.[28,42] Some of these conclusions have previously been proposed from interpretations of chromatographic studies and from theoretical models of the bonded phase surface. However, conclusions which have been drawn from NMR and other types of spectroscopic studies provide concrete and specific supporting evidence for prior chromatographic and theoretical models.

6 ACKNOWLEDGEMENTS

The authors thank Antony J. Williams and the University of Ottawa for assistance with the ^{13}C NMR measurements on LT1. Jim Breeyear's technical assistance with the remaining NMR experiments is greatly appreciated. The authors also thank Charles H. Lochmüller for providing the Partisil silica for the LT1 stationary phase from the Duke Standard Collection established by a grant from Whatman-Chemical Separations. Grateful acknowledgement is made to the University Committee on Research and Scholarship at the University of Vermont, the Society for Analytical Chemists of Pittsburgh and to the donors of the Petroleum Research Fund, administered by the ACS, for partial financial support of this research.

7 REFERENCES

1. J.A. Marqusee and K.A. Dill, *J. Chem. Phys.*, 1986, 85, 434.
2. K.A. Dill, *J. Phys. Chem.*, 1987, 91, 1980.
3. J.G. Dorsey and K.A. Dill, *Chem. Rev.*, 1989, 89, 331.
4. D.E. Martire and R.E. Boehm, *J. Phys. Chem.*, 1983, 87, 1045.
5. C.H. Lochmüller and M.L. Hunnicutt, *J. Phys. Chem.*, 1986, 90, 4318.
6. K.B. Sentell and J.G. Dorsey, *Anal. Chem.*, 1989, 61, 2373.

7.   K.B. Sentell and J.G. Dorsey, J. Chromatogr., 1989, 461, 193.
8.   K.B. Sentell and A.N. Henderson, Anal. Chim. Acta, 1991, 246, 139.
9.   B.C. Gerstein, Anal. Chem., 1983, 55, 781A.
10.   K.B. Sentell, J. Chromatogr., 1993, in press.
11.   R.K. Harris, "Nuclear Magnetic Resonance Spectroscopy: A Physicochemical View", Pitman Books, London, 1983.
12.   A.D. Bain, J. Magn. Reson. 1990, 89, 153.
13.   K.B. Sentell, K.W. Barnes and J.G. Dorsey, J. Chromatogr., 1988, 455, 95.
14.   K. Albert, B. Evers, and E. Bayer, J. Magn. Reson. 1985, 62, 428.
15.   D.B. Marshall and W.P. McKenna, Anal. Chem., 1984, 56, 2090.
16.   D.M. Bliesner and K.B. Sentell, J. Chromatogr., 1993, 631, 23.
17.   D.M. Bliesner and K.B. Sentell, Anal. Chem., 1993, 65 (14), in press.
18.   J.A. Glasel and K.H. Lee, J. Am. Chem. Soc., 1974, 96, 970.
19.   E.H. Ellison and D.B Marshall, Anal. Chem., 1991, 95, 808.
20.   D.E. Woessner and B.S. Snowden Jr., J. Colloid and Interface Sci., 1970, 34, 290.
21.   R.M. McCormick and B.L. Karger, Anal. Chem., 1980, 52, 2249.
22.   E.D. Katz, K. Ogan and R.P.W. Scott, J. Chromatogr., 1986, 352, 67.
23.   E.D. Katz, C.H. Löchmuller and R.P.W. Scott, Anal. Chem., 1989, 61, 349.
24.   M.E. Montgomery Jr., M.A. Green and M.J. Wirth, Anal. Chem., 1992, 64, 1170.
25.   R.M. McCormick and B.L. Karger, J. Chromatogr., 1980, 199, 259.
26.   C.R. Yonker, T.A. Zwier and M.F. Burke, J. Chromatogr., 1982, 241, 257.
27.   C.R. Yonker, T.A. Zwier and M.F. Burke, J. Chromatogr., 1982, 241, 269.
28.   D.M. Bliesner, Ph.D. dissertation, University of Vermont, 1992.
29.   A. Alvarez-Zedpeda, B.N. Barman and D.E. Martire, Anal. Chem., 1992, 64, 1978.
30.   Y. Marcus and Y. Migron, J. Phys. Chem., 1991, 95, 400.
31.   K.L. Rowlen and J.M. Harris, Anal. Chem., 1991, 63, 964.
32.   L.A. Cole and J.G. Dorsey, Anal. Chem., 1990, 62, 16.
33.   K.T. Gillen, M. Schwartz and J.H. Noggle, Mol. Phys., 1971, 20, 899.
34.   E. Breitmaier, K.H. Spohn and S. Berger, Angew. Chem. Int. Ed., 1975, 14, 144.
35.   J.B. Lambert, R.J. Nienhuis and J.W. Keepers, Angew. Chem. Int. Ed., 1981, 20, 487.
36.   R.K. Gilpin and M.E. Gangoda, Anal. Chem., 1984, 56, 1470.
37.   D.W. Sindorf and G.E. Maciel, J. Am. Chem. Soc., 1983, 105, 1848.
38.   E. Bayer, A. Paulus, B. Peters, G. Laupp, J. Reiners, and A. Klaus, J. Chromatogr., 1986, 364, 25.
39.   M. Gangoda, R.K. Gilpin and J. Figueirinhas, J. Phys. Chem., 1989, 93, 4815.
40.   R.C. Zeigler and G.E. Maciel, J. Phys. Chem., 1991, 95, 7345.
41.   E.C. Kelusky and C.A. Fyfe, J. Am. Chem. Soc., 1986, 108, 1746.
42.   D.M. Bliesner, S.T. Shearer and K.B. Sentell, in preparation for Anal. Chem.

Spectroscopic and Chromatographic Characterization of a Self-Assembled Monolayer as a Stationary Phase

M. J. Wirth and H. O. Fatunmbi
DEPARTMENT OF CHEMISTRY & BIOCHEMISTRY, UNIVERSITY OF DELAWARE, NEWARK, DE 19716, USA

1 INTRODUCTION

Recently self-assembled trifunctional silane monolayers have been introduced as a new bonding scheme for chromatographic stationary phases [1-3]. These self-assembled monolayers serve to block the silica substrate from access to the mobile phase, thus increasing the hydrolytic stability at pH extremes and diminishing the adsorption of organic amines. This paper describes the principles of self-assembled trifunctional silane monolayers, their chromatographic performance, and the characterization of their structures by solid state NMR spectroscopy.

The bonding of trifunctional silanes to surfaces is called self-assembly when the bonding density of functional groups is made to be close-packed, approximately 8 μmol/m^2 (22 Å^2/chain) [4-7]. Conventional polymeric phases used in chromatography are no more than 5.5 μmol/m^2 [8]. The difference between these two types of trifunctional silane structures is illustrated by the idealized drawings in Figure 1. The self-assembled monolayer, shown on the left, has alkyl chains that are close-packed, while the conventional polymeric phase, shown on the right, does not. The high density of the self-assembled monolayer offers the advantage of sterically protecting the silica substrate. These two trifunctional silane phases differ due to the differences in the control of the amount of water in the polymerization processes. The conventional polymeric phase is synthesized using a great

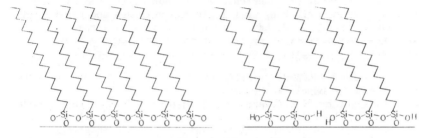

Figure 1. Schematic drawings of self-assembled monolayer (left) and conventional polymeric monolayer.

excess of water, presumably forming a large amount of oligomers
during the bonding process. The self-assembled monolayer is made
by using only the intrinsic monolayer of water on the silica
surface, which provides little, if any, excess water, and the
water is located at the place where the attachment to the surface
occurs. Silica is known to adsorb a monolayer of water over a
wide range of humidities [9]. When reagent silanes react with
the water, they are already in the preferred geometry for surface
attachment. The control of water is thus the key difference
between self-assembly of trifunctional silanes and conventional
polymerization of trifunctional silanes.

For the pure C_{18} self-assembled monolayer, the density is so high
that this monolayer may be of limited use as a chromatographic
phase. Yet the high density is the desired property. To take
advantage of self-assembly in chromatography, mixed phases are
used. Mixed self-assembled monolayers [1-3], illustrated in
Figure 2, combine the advantage of high bonding density with the
ability to control the C_{18} coverage to any desired amount. The
monolayer illustrated in Figure 2 has a long chain functional
group, e.g., C_{18}, and a short chain spacer group, e.g., C_3.

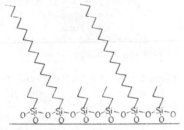

Figure 2. Schematic drawing of the mixed self-assembled
monolayer.

The mixed self-assembly process is applicable to virtually any
type of functional group. In addition to C_{18} groups, we have
used C_8 and C_4 groups, as well as diol groups. Commercially
available trifunctional silanes are available for a whole host of
functional groups, such as amino, cyano, hydroxyl,
polyethyleneimine, allyl, vinyl and epoxide. These last three
groups are particularly useful, as they allow chemical attachment
of a variety of other functional groups not easily bonded through
silane chemistry, such as chiral groups, dextrose and other
hydrophilic groups. We have primarily used C_3 as a spacer and we
are presently investigating the use of C_1 and other spacers.
Self-assembly of trifunctional silanes is widely applicable.

2 CHROMATOGRAPHIC PERFORMANCE

Self-assembled monolayers of trifunctional silanes were prepared
on silica gel by first cleaning and drying the silica gel, and
then passing humidified nitrogen through the silica gel until the
humidity after the silica gel reached that of the entering
nitrogen stream [3]. The humidified silica gel was then poured
into a solution of mixed trichlorosilanes in n-heptane and
allowed to react for one day. After cleaning and drying, the
bonded silica gel was packed into chromatographic columns. The

composition and chromatographic performance were reproducible. Most applications of HPLC involve monomeric rather than polymeric phases because polymeric phases are reputed to be irreproducible. Monomeric phases are prepared by using chlorodimethyl-octadecylsilane as the reagent, and the phase is called monomeric because there must be 1:1 attachment between the reagent groups and the silica gel. The bonding densities of commercially available monomeric phases are typically on the order of 3 $\mu mol/m^2$. The density of available silanols on silica gel is 8 $\mu mol/m^2$ [10], so there are many so-called residual silanols on a typical monomeric bonded phase. It is quite common to use a process called endcapping to reduce the number of residual silanols. This is done by reacting the bonded phase with chlorotrimethylsilane, which is a small reagent and, therefore, is able to reach silanols that the larger silanizing reagents cannot reach. The chromatographic performance of the self-assembled monolayers will be compared with that of conventional endcapped monomeric phases.

A comparison of chromatographic separations by a self-assembled C_8/C_1 phase and a conventional endcapped C_8 monomeric phase (3 $\mu mol/m^2$) is illustrated in Figure 3. The solutes separated (in order of increasing retention time) are uracil, aniline, hexanophenone, hexylbenzene and benzo(a)pyrene, and the mobile phase was 80% acetonitrile/water at 50° C (neutral pH). The chromatogram on the left half of Figure 3 was obtained for the mixed phase, packed in our laboratory without optimization, and chromatogram on the right hand side is that for a commercial C_8 column obtained from Whatman. Both used 10 μm Partisil silica gel. The comparison shows that generally the same behavior is obtained for the two different bonding schemes. The C_8 coverage of the self-assembled monolayer was not characterized and may account for the shorter retention times. In fact, when the same coverages are used for C_{18}, virtually the same retention times are obtained [3].

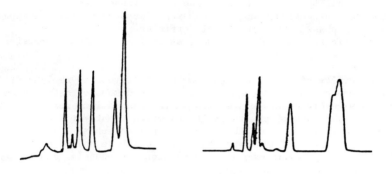

Figure 3. Chromatograms obtained for the mixed self-assembled C_8/C_1 phase (left) and an endcapped monomeric C_8 phase (right).

The chromatogram of the same solutes under the same conditions was obtained for a professionally packed C_{18}/C_3 self-assembled phase (3 μmol/m^2 C_{18}), although the packing procedure was unoptimized. The chromatogram is shown in Figure 4. In this case, 5 μm Zorbax RX was used as the silica substrate. The chromatogram shows that separation efficiencies as high as those for commercial monomeric phases are obtained for these solutes.

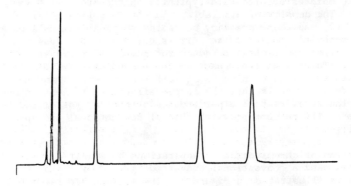

Figure 4. Chromatogram obtained for the same set of test solutes using a professional packed mixed C_{18}/C_3 self-assembled phase.

An important and difficult challenge in chromatography is the separation of peptides and proteins. It is well known that the amino terminus of peptides and proteins has a very high affinity for the residual silanols of silica gel. These silanols are acidic, and it is actually the deprotonated silanols that attract the protonated amino groups at neutral pH. The endcapping of conventional monomeric phases provides some improvement in the performance of these separations, and the preparation of silica gels with less acidity provides additional improvements. Performing the separations at low pH (e.g., pH=2) often allows excellent separations by protonating the silanols, although the conventional phases quickly degrade under these conditions. The mixed self-assembled monolayer behaves like a very well endcapped monomeric phase, providing highly efficient separations for hydrophobic and cationic peptides, as well as difficult proteins [11].

The stability of C_{18}/C_3 self-assembled monolayers (3 μmol/m^2 C_{18}) at extreme pH has been compared to endcapped monomeric C_{18} phases of the same C_{18} coverage prepared on the same silica gel [3]. At both low pH and high pH, the self-assembled monolayer is considerably more stable than the conventional phase. The self-assembled monolayer is virtually inert to mobile phases of pH=2, which offers great promise for separations of proteins and peptides.

3 NMR CHARACTERIZATION
The coverage of C_{18} in the mixed C_{18}/C_3 monolayer has been referred to several times in this paper without noting how it was measured. Quantitative ^{13}C NMR spectroscopy, using cross-

polarization from the alkyl protons, and magic angle spinning, was used to determine the ratio of C_{18} to C_3 groups in the monolayer [12]. Using this ratio, the total coverage was characterized by microanalysis of total carbon content. The of C_{18} and C_3 are also determined in the course of this analysis. These measurements also revealed the individual coverages of C_{18} and C_3. The total coverage of 8 $\mu mol/m^2$ agrees with the coverage that one would expect for close-packed alkyl chains, based on spectroscopic measurements of C_{18} self-assembled monolayers prepared on silicon wafers [4-7]. Incidentally, it is just a coincidence that the intrinsic coverage of silanols on silica is the same 8 $\mu mol/m^2$ as that of the self-assembled monolayer, and it does not mean that there must be 1:1 attachment of reagents to silanol sites.

Another characteristic of the mixed self-assembled monolayer that is shown by the ^{13}C NMR spectral data is that the C_{18} and C_3 groups appear to be randomly intermixed in the monolayer. This information comes from the width of the band that corresponds to most of the carbons in the chain backbone: the width is narrow [12]. Monomeric phases have narrow bands for these backbone carbons, while polymeric phases have wide bands. The bandwidth of the self-assembled phase is similar to that of the monomeric phase. Physically, the bandwidth is a measure of how hindered the motions of the chains are: the polymeric chains are very hindered because the C_{18} chains are packed so tightly. If the C_{18} chains of the self-assembled monolayer had aggregated together on the surface, a wide band would have been observed for the mixed self-assembled phase. It is not surprising that the C_{18} and C_3 groups are randomly intermixed: with n-heptane as the solvent used in the preparation, interactions between C_{18} chains are not much more important that solvent-C_{18} interactions. The mixed C_{18}/C_3 self-assembled monolayer thus behaves like a conventional monomeric phase of the same C_{18} coverage because the chains are spaced from one another like those of the monomeric phase.

The bonding of the reagent groups in the self-assembled monolayer was characterized by ^{29}Si NMR, with cross-polarization from protons and magic angle spinning [12]. ^{29}Si NMR with cross-polarization is an excellent way of obtaining surface sensitivity [13-15]. Surface sensitivity is essential because there is considerably more silicon in the silica gel substrate than there is in the surface monolayer. The bulk silica gel has little available proton intensity to allow cross-polarization, hence its signal is weak.

^{29}Si NMR provides structural information by virtue of its ability to distinguish different chemical types of reagent silicon atoms in the monolayer. There are three possible types of silicon atoms in the self-assembled monolayer: reagent siloxane groups, reagent silanol groups and reagent geminal silanol groups. A perfect self-assembled monolayer would have each reagent group attached to other reagent groups through all three of its siloxane bonds, and this type of group has the formula $RSi(O-)_3$, where R represents the alkyl chain. This form is the reagent siloxane group. In practice, only about half of the groups are

reagent siloxane groups, and the other half are almost all
reagent silanol groups, which have the formula $RSi(O-)_2OH$.
Further, about half of the reagent siloxane groups involve
attachment of one of the three bonds to a surface silanol group.
This was determined by using quantitative measurements from
sample-to-sample to determine how many substrate silanol groups
were consumed in the monolayer formation. Finally, there are
very few reagent geminal silanol groups, which have the formula
$RSi(O-)(OH)_2$. With so little reagent siloxane groups, the self-
assembled monolayer is far from perfect, yet it still has the
high 8 $\mu mol/m^2$ density. A mathematical analysis of the NMR data
revealed that the self-assembled monolayer consisted of long
chains that were occasionally cross-linked to another long chain.
The 8 $\mu mol/m^2$ coverage indicates that these polymeric chains lie
close to one another.

There is a very good reason why the polymerization of the self-
assembled monolayer is linear rather than two-dimensional: it is
sterically impossible for C_{18} and C_3 chains to polymerize in two
dimensions in a plane. The distance between silicon atoms of two
reagent groups bonded together through a siloxane linkage is
determined by the Si-O-Si and O-Si-O bond angles and the Si-O
bond distance. The spacing between adjacent reagent silicon
atoms is not wide enough to accommodate the alkyl groups, given
the known bond angles and bond distances of model compounds and
the van der Waals radii of the hydrogens on the alkyl groups. An
accommodation is possible only for two linked reagent adjacent
groups because the third group (either an OH or a substrate
attachment) can be directed toward the surface (i.e., it goes out
of the plane). We are presently exploring other types of spacer
groups that might allow for two-dimensional siloxane
polymerization. Such an accomplishment could have a profound
impact on separation science by virtue of its ultimate blockage
of the silica substrate.

4 FUTURE DIRECTIONS

The ability to make more stable self-assembled monolayers and to
passivate the underlying silica substrate is important to a range
of applications in separation science owing to the wide use of
silica. Studies are underway for separating peptides and
proteins by HPLC, as well as basic drugs. In addition to these
chromatographic studies, we are exploring the application of
monolayer self-assembly to capillary zone electrophoresis.
Bonding of surface groups by monolayer self-assembly would
greatly benefit capillary zone electrophoresis, where silica is
the material of which the capillaries are made. Electro-osmotic
flow, which is detrimental to separations, is caused by ionizable
groups on the capillary walls.   Self-assembly would
make these groups unable to ionize. Further, large proteins can
irreversibly adsorb to the capillary walls unless very
hydrophilic groups such as sugars or polyacrylamide are used
[16]. Such groups can presumably be attached to the self-
assembled monolayer to provide a hydrophilic and hydrolytically
stable coating over the hydrophilic wall.

ACKNOWLEDGEMENTS
This work was supported by the Department of Energy under contract DE-FG02-91ER14187.

REFERENCES
1. M.J. Wirth and H.O. Fatunmbi, U.S. Patent Application Number 900,215, filed on June 17, 1992.
2. M. J. Wirth and H. O. Fatunmbi, *Anal. Chem.*, 1992, <u>64</u>, 2783.
3. M. J. Wirth and H. O. Fatunmbi, *Anal. Chem.*, 1993, <u>65</u>, 822.
4. Moaz, R.; Sagiv, J., *J. Colloid. Int. Sci.*, 1984, <u>100</u>, 465.
5. Moaz, R.; Sagiv, J., *Langmuir*, 1987, <u>3</u>, 1034.
6. Wasserman, S.R.; Tao, Y.-T.; Whitesides, G.M., *Langmuir*, 1989, <u>5</u>, 1074.
7. Wasserman, S.R.; Whitesides, G.M.; Tidswell, I.M.; Ocko, B.M.; Pershan, P.S.; Axe, J.D., *J. Am. Chem. Soc.* 1989, <u>111</u>, 5852.
8. Sander, L.C.; Wise, S.A., *Anal. Chem.*, 1984, <u>56</u>, 504.
9. Gee, M.L.; Healy, T.W.; White, L. R., *J. Colloid. Int. Sci.*, 1990, <u>140</u>, 450.
10. Zhuravlev, L.T. *Langmuir*, 1987, <u>3</u>, 316.
11. M.J. Wirth and H.O. Fatunmbi, to be published.
12. H.O. Fatunmbi, M.D. Bruch and M.J. Wirth, *Anal. Chem.*, in press.
13. Maciel, G.E.; Sindorf, D.W. *J. Am. Chem. Soc.* 1980, <u>102</u>, 7606.
14. Sindorf, D.W.; Maciel, G.E. *J. Am. Chem. Soc.* 1981, <u>103</u>, 4263.
15. Sindorf, D.W.; Maciel, G.E. *J. Am. Chem. Soc.* 1983, <u>105</u>, 3767.
16. Cobb, K.A.; Dolnik, V.; Novotny, M., *Anal. Chem.*, 1990, <u>62</u>, 2478.

FTIR Study of Adsorption at the Silica/Solution Interface: Interaction of Surface Sites with Carbonyl Groups

Jean-Marc Berquier

LABORATOIRE CNRS/SAINT-GOBAIN, 'SURFACE DU VERRE ET INTERFACES', 39, QUAI LUCIEN LEFRANC, 93303 AUBERVILLIERS, FRANCE

1 INTRODUCTION

We have developed in our laboratory an attenuated total reflection (ATR) experiment which permits to study in situ and with a very high sensitivity, the interactions of a flat silica surface with different organic products (e.g. silanes, polymers, ...). A description of this experiment and the first results concerning silane reaction with the surface have already been published[1]. In this present work, we study the adsorption of two organic species on the silica surface sites. Both comprise carbonyl groups which are able to interact through hydrogen-bond with the silica silanols while the adsorption proceeds. We report the spectra associated with vibrations of functional groups involved in these hydrogen-bond interactions. Firstly, the adsorption of acetone from carbon tetrachloride is studied. We use this example to show how quantitative information may be obtained by assuming a model for the adsorption isotherm : the strength of the interaction and the number of the sites involved are determined. Secondly, preliminary results concerning the adsorption of polymethylmethacrylate (PMMA) on silica and silanized silica are presented. We compare this process with the acetone adsorption, with particular interest for the surface sites in interaction.

2 EXPERIMENTAL PROCEDURE

Infrared Spectroscopy

A Nicolet 800 FT-IR spectrometer equipped with a narrow-band mercury-cadmium-telluride (MCT-A) detector has been used. Infrared spectra have been recorded at a resolution of 4 cm^{-1} by coadding 256 interferograms. The ATR attachment is a beam-condenser from Harrick Sc. Corp.. The flow cell designed to fit the optics and to support the crystal uses teflon as the only material in contact with the solution.

Substrate

The crystal is a 50 x 10 x 0.5 mm³, 45° trapezoidal single crystal cut from a silicon wafer. It is covered by the native oxide layer, approximately 20 Å thick, which is prepared using the following chemical treatment : CHCl3 for 10 minutes in an ultrasonic bath, HF-DI water (1:10 by volume) until hydrophobic, DI water rinse, DI water-H2O2-NH4OH (5:1:1 by volume) at 80°C for 10 minutes, DI water rinse, DI water for 10 minutes in an ultrasonic bath, final drying in a N2 gas blow.

Chemical reagents

Carbon tetrachloride is dried over molecular sieves. Acetone is used as received from Merck. Polymethylmethacrylate (PMMA), molecular weight 93000, is from Aldrich. PMMA, molecular weight 3000 and molecular weight distribution Mw/Mn ≤ 1.15, is from Polymer Laboratories Ltd.

ATR FT-IR Experiments

A peristaltic pump and a three-way valve are used to circulate either the pure solvent or the solutions through the flow cell. After the crystal preparation, the flow cell is immediatly assembled and aligned in the purged sample compartment of the spectrometer and the solvent is pumped into the cell. A background spectrum is collected ca. one hour after the cell has been filled with the solvent. Then, the solution is pumped into the cell. For the acetone adsorption experiments the acetone solutions with increasing concentrations are pumped successively. Two minutes after a solution is pumped, a spectrum is recorded. For the PMMA adsorption experiment, the polymer is allowed to adsorb for 1 h. Next the polymer solution is replaced by pure solvent. A spectrum is then collected five minutes after the solvent is pumped into the cell.

3 RESULTS AND DISCUSSION

Acetone Adsorption

Figures 1 and 2 show two frequency ranges of ATR spectra recorded for different solutions of acetone in carbon tetrachloride, ratioed to the spectrum of the cell containing pure CCl4. In Figure 1 two peaks (the intensity of which increases with concentration) appear in the range 1820-1620 cm⁻¹. These bands correspond to the stretching vibration of the C=O group of acetone molecules in two different environments : in the solution and at the surface. If we assume that the interaction of an acetone molecule with the surface occurs through the formation of an hydrogen bond between the carbonyl group and a surface silanol, this will result in a shift of the C=O stretching

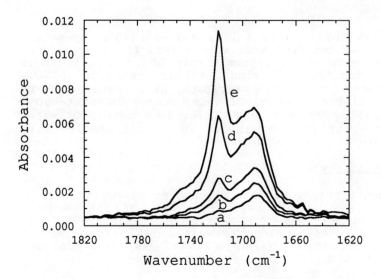

Figure 1 ATR spectra of silica in contact with different concentrations of acetone : a) 3.10^{-4}M, b) 5.10^{-4}M, c) 10^{-3}M, d) 3.10^{-3}M and e) 5.10^{-3}M. The intensity of the C=O bands increases with concentration.

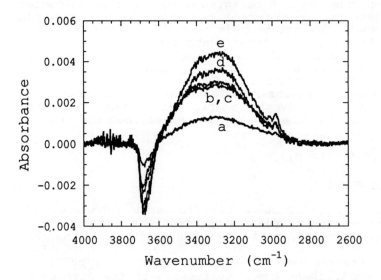

Figure 2 Same as in Figure 1 in the wavenumber range of the stretching vibration of the surface silanols O-H. The intensity of the negative peaks (ca. 3670 cm^{-1}) and of the broad peaks (ca. 3350 cm^{-1}) increases with concentration.

vibration frequency. Accordingly, the bands at 1717 cm⁻¹ and at 1690 cm⁻¹ may be ascribed to the acetone molecules in solution and to the acetone molecules adsorbed on the silica surface, respectively. The observed shift of 27 cm⁻¹ is in agreement with the literature[2]. The formation of the hydrogen-bond will also result in a shift of the silanol O-H stretching vibration frequency, which is illustrated in Figure 2. The negative peak is assigned to the surface sites which were previously in contact with carbon tetrachloride ("free" silanol) : indeed its frequency, in the range 3680 - 3650 cm⁻¹, corresponds to reactive surface silanols in a CCl₄ environment[3]. The positive broad band is assigned to these sites which are now in interaction through hydrogen-bonding with acetone molecules (H-bonded silanol).

In order to obtain more quantitative information from the spectra, we decompose with a computer program the peaks in the range 1820-1620 cm⁻¹. The intensity of the two bands of the C=O group is given in Figure 3 as a function of the nominal concentration. At this point, it is important to notice that the intensity of the solution acetone band varies linearly with the concentration. The slope of this line will be used later.

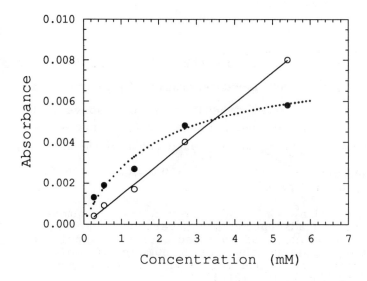

Figure 3 Intensity of the two carbonyl bands at 1690 cm⁻¹ (closed marks) and at 1717 cm⁻¹ (open marks) as a function of acetone concentration. The solid line is the linear least-squares fit of the data at 1717 cm⁻¹. The dotted line is a guide for the eyes.

The adsorption of acetone occurs essentially through hydrogen-bonds with the surface silanols : all adsorbed species only interact with a surface site, therefore, adsorption is limited to a monolayer. The interaction between adsorbed molecules is not relevant. Furthermore, because the heat of adsorption through hydrogen-bond is fairly small in comparison with chemical bond energies[4], the adsorption may be seen as a reversible process and the adsorbed molecules may be considered to be in equilibrium with the molecules in solution, which we experimentally verified. For all these reasons, we will assume that this adsorption isotherm is of Langmuir type. In this framework, we will define the algebraic formulation of the model isotherm function[5] and will try to fit the variation of the intensity of the H-bonded carbonyl band.

The model treats the adsorption process as a competition between solute and solvent :

$$(acetone)_{sol.} + (CCl_4)_{surf.} \Leftrightarrow (acetone)_{surf.} + (CCl_4)_{sol.} \quad (1)$$
$$\phantom{(acetone)_{sol.}} c_2 N_1^s N_2^s a_1$$

with an equilibrium constant K,

$$K = \frac{N_2^s \cdot a_1}{N_1^s \cdot c_2} \quad (2)$$

where $N_1^s + N_2^s = 1$, a_1 is the solvent activity in solution, c_2 is the acetone concentration in solution, and N_1^s and N_2^s are the mole fractions of the solvent and acetone in the adsorbed layer, respectively. Since we only used dilute solutions, c_2 has replaced the acetone activity in solution and a_1 can be considered as constant. Conveniently we define $b = K / a_1$.

Introducing $n_2^s = N_2^s \cdot n^s$, where n_2^s is the number of moles of occupied adsorption sites per square meter, and also the number of moles of adsorbed acetone molecules per square meter, and n^s the total number of moles of adsorption sites per square meter, the following equation is obtained from equation (2) :

$$\frac{c_2}{n_2^s} = \frac{1}{n^s \cdot b} + \frac{c_2}{n^s} \quad (3)$$

which allows the determination of n^s and b through the plot of c_2/n_2^s versus c_2.

The values of c_2 and n_2^s which are used in our calculation, are given by the intensity of the bands at 1717 and 1690 cm^{-1}. The relations between these values and the intensities will not be described here but a detailed discussion for similar relations can be found elsewhere[1]. However it should be pointed out that the calculation of these relations is based on the assumption that the vibration modes of the grafted molecules remain anisotropic and that the infrared absorptivity of the carbonyl group remains the same in the two environments. We are aware that these assumptions are certainly rough. For example, it is known that the absorptivity of the hydrogen-bonded carbonyl may be slightly enhanced[6]. We expect that the exact value of the calculated constants will be affected but that the main conclusions of the analysis emphasized here will remain unchanged.

We have used the value of the slope of the linear relation (see Figure 3) and the value of the penetration depth[7] at 1700 cm^{-1} for the calculation. We obtain these values of n^s and b :

$n^s = 1.3 \ 10^{-6}$ mole/m^2 and b = 593 mole^{-1} .

The value of n^s corresponds to a silanol density of 0.7 nm^{-2}, which is lower than expected from the literature[8] but in a very good agreement with the value of 0.65 nm^{-2} obtained from the grafting of a monohydrolyzable silane[1].

From b, using a value of a_1 = 10.36 moles/liter, we find K = 6146. From this value, $\Delta G°$ for the adsorption equilibrium at 293 K is tentatively estimated to be equal to 5.1 kcal/mole. This value is typical of the formation of a hydrogen-bond[4,9].

PMMA Adsorption

During the PMMA adsorption process, some ester groups of the adsorbed polymer interact with the surface sites through hydrogen-bonds[10-12]. As well as for the acetone adsorption, the ATR spectrum after the adsorption will give some information on the interacting polymer and surface sites.

The 1800-1650 cm^{-1} range of the ATR spectrum is presented in Figure 4 in comparison with a transmittance spectrum of the carbonyl peak of PMMA in CCl$_4$. It is clear that the ATR spectrum is asymmetric with a shoulder at ca. 1716 cm^{-1}, which may be ascribed to the hydrogen-bonded carbonyl vibration, in good accordance with literature[10-12]. In order to illustrate this effect, the transmittance spectrum has been substracted from the ATR spectrum, yielding the remainder also shown in Figure 4.

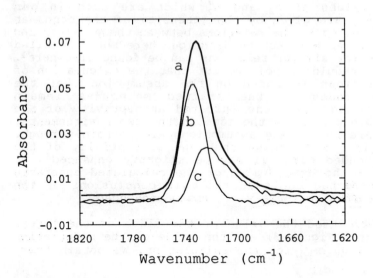

Figure 4 a) ATR spectrum of adsorbed PMMA on silica in the region of carbonyl absorption. b) liquid transmission spectrum. c) a - b.

In Figure 5, the spectrum of the adsorbed PMMA is compared with the spectrum obtained during the acetone adsorption. Similar negative peaks at about 3660 cm^{-1} are found in both spectra but, in contrast to the acetone adsorption, two positive peaks appear in this range at 3400 and 3250 cm^{-1} for the PMMA adsorption. This would mean that the surface sites interact with the PMMA in two different manners, which we ascribe to the interactions with the two different oxygen elements of the ester group. A consequence would be that, as the shoulder due to the hydrogen-bonded carbonyls is only representative of one interaction type, the amount of interacting polymer segments could be higher than the value measured from the hydrogen-bonded carbonyl band.

PMMA Adsorption on a silanized silica surface

In order to see if it is the same type of surface sites, corresponding to the vibration band at about 3670 cm^{-1}, which interact with silanes or with the polymer, we have studied the adsorption of PMMA on silanized surfaces with different silane coverages. HMDS (hexamethyldisilazane) which is commonly used for deactivating chromatographic support materials[13,14], has been chosen to treat in situ the silica surfaces. We have measured the intensity of the silanol band after the silanization and after the following adsorption, for two experiments which only differ by the parameters of the silane reaction. The experiment 1 used a longer reaction

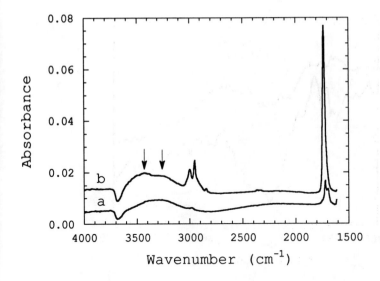

Figure 5 Comparison between the spectra after adsorption of a) acetone and b) PMMA. Note the two broad bands in spectrum b between 3500 and 3200 cm^{-1}.

time (17 hours) and a higher HMDS concentration (10^{-2}M) than the experiment 2 (50 minutes and 10^{-3}M, respectively). The spectra in the 3800-3500 cm^{-1} range corresponding to the first experiment, are shown in Figure 6. The values of the intensity of the band corresponding to the surface sites can be found in Table 1. In the two experiments, the total intensity is similar. This confirms that is the same type of surface sites, corresponding to this band, which is able to react to silane or to interact through hydrogen-bond.

Table 1 Intensity of the band (ca. 3670 cm^{-1}) for the different steps of the experiments 1 and 2. (Absorbance unit)

	Experiment 1	Experiment 2
Silanization	-16 ± 3.10^{-4}	-8 ± 3.10^{-4}
Adsorption	-9 ± 3.10^{-4}	-16 ± 3.10^{-4}
Total	-25 ± 3.10^{-4}	-24 ± 3.10^{-4}

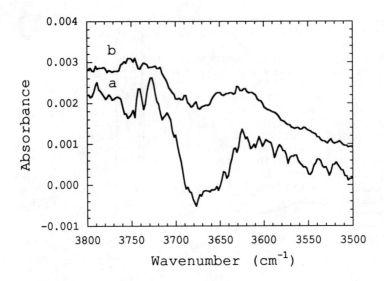

Figure 6 ATR spectra, in the frequency range of the stretching vibration of the surface silanols, of the silica a) after partial silanization by HMDS (background spectrum : pure solvent before silanization). b) after silanization and adsorption of PMMA. (background spectrum : spectrum a)

4. CONCLUSION

We have shown that the adsorption of organic products at the silica surface can be studied in situ with a very high sensitivity by this ATR technique.

Qualitative and quantitative information on the interactions of adsorbed acetone molecules with the surface sites are obtained. Furthermore we have shown that the amount of available surface sites can be estimated through the measurement of the adsorption isotherm. These sites are able to react with silane as well as to interact with polymer segments. In the case of PMMA adsorption, two types of interaction between the surface sites and the ester groups are revealed.

REFERENCES

1. M.J.Azzopardi and H.Arribart, J.Adhesion, in press.
2. D.M.Griffiths, K.Marshall and C. H. Rochester, J.C.S.Faraday I, 1974, 70, 400.
3. W.D.Bascom, J.Phys.Chem., 1972, 76, 3188.
4. N.L.Allinger et al,'Chimie Organique', McGraw-Hill, Paris, 1982, p 92 de la version française.

5. A.W.Adamson, 'Physical Chemistry of Surfaces', Wiley-
    Intersciences , 4th Ed. 1982, chapter XI.
6. M.Coleman and P.Painter, Appl.Spectrosc.Rev.,
    1985, 20, 255
7. F.M.Mirabella Jr and N.J.Harrick, 'Internal reflection
    spectroscopy : review and supplement',
    Marcel Dekker Ed, 1985.
8. M.L.Hair, 'IR Spectroscopy in Surface Chemistry',
    Marcel Dekker Ed, NY, 1967
9. H.Knözinger, 'The hydrogen bond - recent development
    in theory and experiments', P.Schuster et al Eds,
    North-Holland Publ.Co, Amsterdam, 1976, Chapter 27.
10. B.J.Fontana and J.R.Thomas,
    J.Phys.Chem., 1961, 65,480.
11. D.L.Allara, Z.Wang and C.G.Pantano,
    J.Non-Cryst.Solids,1990,120,93
12. H.E.Johnson and S.Granick, Macromol., 1990, 23, 3367.
13. W.Hertl and M.L.Hair, J.Phys.Chem.,1971,75,2181
14. C.P.Tripp and M.L.Hair, in 'Chemically Modified Oxide
    Surfaces', D.Leyden and W.Collins Ed.,
    Gordon and Breach Sc.Pub., 1990, 375.

Subject Index